Selection of Deep-Sea Aquaculture Species and Development of Supporting Technologies and Equipment

Selection of Deep-Sea Aquaculture Species and Development of Supporting Technologies and Equipment

Guest Editors

Zhenhua Ma
Jianguang Qin

MDPI

Basel • Beijing • Wuhan • Barcelona • Belgrade • Novi Sad • Cluj • Manchester

Guest Editors
Zhenhua Ma
South China Sea Fisheries Research Institute
Chinese Academy of Fishery Sciences
Guangzhou
China

Jianguang Qin
Southern Marine Science and Engineering Guangdong Laboratory
Guangzhou
China

Editorial Office
MDPI AG
Grosspeteranlage 5
4052 Basel, Switzerland

This is a reprint of the Special Issue, published open access by the journal *Journal of Marine Science and Engineering* (ISSN 2077-1312), freely accessible at: https://www.mdpi.com/journal/jmse/special_issues/02IB7R7325.

For citation purposes, cite each article independently as indicated on the article page online and as indicated below:

Lastname, A.A.; Lastname, B.B. Article Title. *Journal Name* **Year**, *Volume Number*, Page Range.

ISBN 978-3-7258-8387-5 (Hbk)
ISBN 978-3-7258-8388-2 (PDF)
https://doi.org/10.3390/books978-3-7258-8388-2

Cover image courtesy of Zhenhua Ma

Contents

About the Editors

Zhenhua Ma

Zhenhua Ma is a Senior Researcher and Vice Director of the South China Sea Fisheries Research Institute, Chinese Academy of Fishery Sciences, and President of the Sanya Tropical Fisheries Research Institute. He received his Ph.D. in Aquaculture from Flinders University, Australia, in 2012, and has interdisciplinary expertise in both aquaculture and information technology. He also serves as an Adjunct Associate Professor at Flinders University and as a doctoral supervisor at several universities in China and Australia. Zhenhua Ma is the Chief Expert of the National Innovation Team for Deep-Sea Aquaculture Technology and Species Development of the Chinese Academy of Fishery Sciences. His research focuses on offshore and deep-sea aquaculture, aquaculture technology and species development, aquatic animal stress physiology, and the application of interdisciplinary approaches to sustainable marine aquaculture. He has extensive experience in research leadership, platform development, and innovation management, and has played a key role in building open collaborative research platforms that connect universities, research institutes, and industry partners. To date, he has led or participated in more than 40 research and technical service projects with total funding exceeding RMB 50 million, received multiple scientific and technological awards, published more than 220 academic papers including over 120 SCI-indexed articles, and edited or co-edited numerous academic books in both English and Chinese. Through his academic leadership, broad international perspective, and long-term commitment to innovation in marine aquaculture, Zhenhua Ma has become a highly respected scholar in the field.

Jianguang Qin

Jianguang Qin is a professor at Guangzhou Marine Laboratory and an adjunct professor at Flinders University, Australia. Jianguang obtained a bachelor's degree at Dalian Fisheries College, China in 1982, a master's degree at the University of Oklahoma in 1988, and a PhD at Ohio State University in 1994. After a three-year postdoctoral position at the University of Hawaii, he joined Flinders University in 1997. Jian is a pioneer of the aquaculture program at Flinders and has led its aquaculture research to world-class standards in Fisheries Sciences. Jianguang Qin has supervised 22 PhD students and 20+ Honors students to completion, and published 4 books, 19 book chapters, and 450+ journal papers. His work has been cited over 18,000 times, and he has an h-index of 68. Jian's research areas cover food, feeding and nutrition of aquatic animals to discover trophic relationships in the food web and apply new findings to applied science. Research disciplines include ecology, physiology, behaviour, biotechnology, and toxicology. Research organisms include bacteria, algae, zooplankton, fish, oyster, abalone and freshwater crayfish. Jianguang Qin was named one of the world's top 100 high-impact scientists in Fisheries in 2022, mentioned among Australia's top research leaders in Marine Science and Fisheries in 2024, and included in Stanford University's list of the world's top 2% scientists over the past five years.

Journal of
*Marine Science
and Engineering*

MDPI

Editorial

Selection of Deep-Sea Aquaculture Species and Development of Supporting Technologies and Equipment

Zhenhua Ma [1,*] and Jianguang Qin [2,*]

1 Sanya Fisheries Research Institute, South China Sea Fisheries Research Institute, Chinese Academy of Fishery Sciences, Sanya 572018, China
2 Southern Marine Science and Engineering Guangdong Laboratory (Guangzhou), Guangzhou 511458, China
* Correspondence: zhenhua.ma@hotmail.com (Z.M.); qin_jianguang@gmlab.ac.cn (J.Q.)

Aquaculture has long been a cornerstone of global food production, providing a significant source of seafood to meet the growing demand for animal protein [1–3]. Traditionally, aquaculture has been confined to shallow waters, particularly coastal zones, where conditions are more favorable for farming marine species. These areas offer relatively easy access to essential resources such as feed, water quality monitoring systems, and established methods for harvesting. The proximity to land also facilitates the logistical aspects of the industry, including transportation and processing. However, over time, several limitations of shallow-water aquaculture have become increasingly evident. One of the most pressing issues is the environmental degradation of coastal ecosystems, including water pollution, habitat destruction, and the depletion of natural fish stocks due to overfishing [4,5]. The intensive farming practices in these areas often result in the accumulation of waste, excess nutrients, and antibiotics, which negatively impact surrounding marine environments. As fish populations in shallow waters continue to dwindle, the ecological balance of these areas is further compromised, driving the need for a more sustainable solution. Considering these challenges, the exploration of deeper marine environments has emerged as a viable alternative, with the shift toward deep-sea aquaculture (also considered as offshore aquaculture) marking a significant transformation in the aquaculture industry. As global seafood demand rises and shallow-water ecosystems face increasing stress, the deep sea has become a promising frontier [6,7]. Deep-sea aquaculture, which involves farming marine species in deeper waters beyond the coastal zones, offers numerous potential advantages—it is able to reduce the pressure on near-shore ecosystems, provide new spaces for cultivation, and foster species diversification [8]. However, its success depends on selecting appropriate species, developing innovative systems, and overcoming the numerous technical challenges posed by these environments (Figure 1).

Selecting species suitable for offshore aquaculture is one of the most critical aspects of developing sustainable and efficient systems. The deep-sea environment poses unique challenges, including fluctuating temperatures, extreme pressure, and limited light penetration. Therefore, only species that are naturally adapted to such conditions can be considered for offshore farming. Species must be able to withstand the harsh conditions of deep-sea environments, including strong currents, variable temperatures, and changes in salinity, and those that can tolerate fluctuating environmental factors, such as temperature and salinity gradients, are prime candidates for offshore cultivation. These species may include certain pelagic fish, shellfish, and marine invertebrates, which have evolved to thrive in deeper, more stable waters. Species selected for offshore aquaculture must exhibit high growth rates to ensure efficient production, as the ability to grow rapidly in the confined

 https://doi.org/10.3390/jmse14111004

space of a farm is essential for economic sustainability. This criterion is particularly important in offshore environments, where logistical challenges such as feed transportation and maintenance costs can be high. Species with high feed conversion ratios and low feed costs are preferable in such systems. In addition to environmental tolerance and growth potential, market demand plays a crucial role in species selection. Species that are in high demand for both domestic consumption and export markets are ideal candidates for offshore aquaculture. Popular species in the global market, such as tuna, salmon, and various shellfish, have already seen successful applications in offshore farming. Moreover, disease management is one of the most significant challenges in aquaculture. Offshore environments, especially in deep-sea locations, present unique biosecurity challenges, including limited access for monitoring and treatment. Species selected for offshore aquaculture must demonstrate natural resistance to common diseases and pathogens, as well as adaptability to the biosecurity measures in place. Additionally, selective breeding programs may be necessary to enhance disease resistance and improve overall farm health.

Figure 1. Conceptual overview of deep-sea aquaculture (AUV: autonomous underwater vehicles, ROV: remotely operated vehicles).

To overcome the challenges posed by deep-sea environments, significant advancements in aquaculture systems and technologies have been made (Table 1). These innovations are crucial for enhancing the sustainability, efficiency, and scalability of offshore aquaculture operations. Key technological breakthroughs include improvements in monitoring systems, autonomous feeding systems, environmental control, and the development of specialized equipment designed to function under deep-sea conditions.

https://doi.org/10.3390/jmse14111004

Table 1. Summary of supporting technologies and equipment for deep-sea aquaculture.

Category	Supporting Technologies/Equipment	Main Functions	Value for Readers and Operators
Monitoring and inspection	Sensor networks, underwater cameras, AUVs, ROVs, and real-time telemetry	Collect data on water quality, currents, species health, cage condition, and farm performance.	Reduces dependence on frequent human offshore intervention and supports early warning, maintenance, and operational decisions.
Precision feeding	Autonomous feeding systems, feed dispensers, behavior sensors, and decision algorithms	Adjust feeding schedules and ration levels according to environmental conditions and species demand.	Improves feed conversion, reduces waste, lowers operating costs, and helps maintain consistent nutrition.
Environmental control	Automated filtration, oxygenators, temperature regulation, salinity monitoring, and water-quality sensors	Maintain suitable oxygen, temperature, salinity, and other key conditions for farmed organisms.	Supports survival and growth in variable offshore environments and helps standardize production conditions.
Closed-loop and low-impact systems	Water recirculation, waste capture/treatment, biosecurity barriers, and reduced-chemical management	Minimize water exchange, nutrient discharge, pathogen transfer, and external chemical inputs.	Helps lower the environmental footprint and aligns deep-sea aquaculture with sustainability expectations.
Farm structures and mooring	Deep-water cages, flexible net pens, corrosion-resistant materials, anchors, mooring lines, and anti-biofouling components	Withstand strong currents, pressure, low temperatures, and corrosive seawater while securing farmed species.	Improves structural safety, stock retention, and long-term reliability in harsh deep-sea conditions.
AI-enabled visual analytics	Underwater image enhancement, multi-target fish tracking, NeuroVI-RGB multimodal monitoring, and segmentation models	Improve perception in low-visibility water and detect feeding desire, biofouling, fence breaches, dead fish, and behavior changes.	Turns complex image data into practical management information for intelligent cage monitoring.
Energy and power support	Marine renewable-energy devices such as vertical-axis hydrokinetic turbines and wave/tidal-current energy modules	Harvest local marine energy to support sensors, cameras, communications, and automated equipment.	Can reduce reliance on shore-based power and improve the feasibility of remote offshore farms.
Germplasm and species improvement tools	Cryopreservation, selective breeding, gene-editing approaches, microbiota assessment, and marine biotechnology	Preserve genetic resources, improve stress tolerance and disease resistance, and support species selection.	Links technology development with biological performance, making species selection more actionable.

One of the primary challenges of deep-sea aquaculture is the difficulty of monitoring and managing farms located in remote, deep-sea environments. Traditional methods of monitoring and maintenance, which rely on regular human intervention, are costly and logistically challenging. To address this, the development of autonomous systems has been a key focus in recent research. Autonomous underwater vehicles (AUVs) and remotely operated vehicles (ROVs) are being utilized to gather real-time data on environmental conditions, water quality, and the health of farmed species. These systems can provide critical insights into the farm's performance, reducing the need for costly and time-consuming human interventions. Autonomous feeding systems are also being developed to optimize feed distribution and ensure efficient utilization, relying on sensors and algorithms to determine the optimal feeding schedules based on environmental conditions and the specific needs of the farmed species. By reducing feed waste and ensuring consistent nutrition, autonomous feeding systems improve farm productivity and reduce operational costs.

In deep-sea aquaculture, maintaining optimal conditions for farmed species is crucial for their survival and growth. Environmental control systems are essential for regulating water temperature, oxygen levels, salinity, and other key factors that influence species health. Advanced environmental control systems, including automated water filtration, oxygenators, and temperature regulators, are being developed to ensure that the farmed species are kept within their ideal conditions.

Innovative technologies, such as closed-loop systems, are being explored to create more sustainable environments for deep-sea aquaculture, helping to minimize water usage and reduce the need for external inputs, such as chemicals and antibiotics, which can harm the surrounding ecosystem. Furthermore, these systems are able to assist in reducing the environmental footprint of offshore farms, making them more environmentally friendly and sustainable.

The harsh conditions of deep-sea environments demand the development of specialized equipment that can function reliably at great depths. This includes underwater cages, feeding systems, monitoring tools, and other infrastructure designed to withstand high pressure, low temperatures, and corrosive seawater. Research in materials science has led to the development of durable, corrosion-resistant materials for use in the construction of offshore aquaculture systems.

Underwater cages, which house the farmed species, must be designed to withstand the physical stress caused by strong currents and the pressure of the deep sea. These cages are often constructed from flexible materials that allow for movement in response to environmental forces. Additionally, advanced mooring systems are required to anchor the cages securely to the seafloor, preventing displacement and ensuring the safety of the farmed species.

Despite the technological advancements in offshore aquaculture, several challenges remain in developing sustainable and efficient systems for deep-sea farming [9]. Offshore aquaculture requires substantial investment in infrastructure, equipment, and technology. The high cost of establishing deep-sea farms, combined with ongoing operational expenses associated with monitoring, feeding, and maintenance, can be a significant barrier to the industry growth. However, technological advancements, such as autonomous systems and more efficient environmental control technologies, have the potential to reduce operational costs over time. While offshore aquaculture offers the potential for resource diversification and reduced pressure on coastal ecosystems, it is not without environmental risks. Issues such as nutrient pollution, the escape of farmed species into the wild, and the impact of disease outbreaks on wild populations must be carefully managed. Sustainable practices, such as the use of closed-loop systems and biosecurity measures, are essential for minimizing the environmental impact of offshore aquaculture. The deep-sea aquaculture area is often located in areas that are difficult to access and are subject to complex regulatory frameworks. The establishment of offshore aquaculture farms requires navigating complex legal and environmental regulations, which can vary significantly between countries and regions. Additionally, logistical challenges, such as transportation of feed, equipment, and harvested products, must be addressed to ensure the long-term viability of offshore farms.

Deep-sea aquaculture holds tremendous potential for diversifying seafood production and alleviating pressure on shallow water ecosystems [10]. However, the success of deep-

https://doi.org/10.3390/jmse14111004

sea aquaculture depends on the careful selection of species that can thrive in harsh deep-sea environments, as well as the development of innovative technologies and equipment that can overcome the unique challenges posed by these habitats. Advancements in autonomous systems, environmental control technologies, and specialized equipment are essential for the growth and sustainability of offshore aquaculture. As industry continues to evolve, it is crucial to balance technological development with environmental sustainability to ensure the long-term success of deep-sea farming and contribute to global food security.

The Special Issue, entitled "Selection of Deep-Sea Aquaculture Species and Development of Supporting Technologies and Equipment", aims to identify viable species for deep-sea aquaculture and develop the technological innovations necessary for their sustainable cultivation. This includes establishing species selection criteria, advancing aquaculture systems, and designing specialized equipment suited to the extreme deep-sea environment. While aquaculture has traditionally been confined to shallow waters, recent developments have shifted focus to deeper, more challenging habitats, driven by the need for resource diversification and environmental sustainability. Cutting-edge research is focused on enhancing species adaptability, improving biosecurity, developing autonomous monitoring and feeding systems, and optimizing environmental control technologies for deep-sea conditions. We welcome contributions that present novel species selection methods, technological breakthroughs in aquaculture infrastructure, and strategies for addressing the unique challenges of deep-sea environments, with an emphasis on sustainability and operational efficiency. The research contents and main findings in this Special Issue are as follows:

In contribution 1, Huang et al. investigate the impact of low-temperature stress on the gut microbiota of juvenile yellowfin tuna, a species vital to aquaculture. The research examines three experimental groups exposed to different temperatures: control (30 °C), low temperature (24 °C), and ultra-low temperature (18 °C). After 36 h of cold stress, the composition and diversity of the gut microbiota significantly changed, particularly at the phylum and genus levels. Proteobacteria remained the dominant phylum, but its relative abundance decreased with lower temperatures, while the relative abundance of Firmicutes and Bacteroidota increased, indicating microbial adaptation to cold conditions. Notably, Acinetobacter became the dominant genus in both low- and ultra-low-temperature groups, suggesting its role in metabolic flexibility and survival under cold stress. Additionally, changes in the richness and diversity indices (Ace and Shannon indices) revealed that while microbial richness increased under extreme cold, diversity showed a decline, followed by recovery, reflecting the dynamic adjustments of gut microbiota to cold exposure. These findings highlight the adaptability of yellowfin tuna to cold stress through shifts in gut microbiota, providing valuable insights for improving aquaculture management under varying environmental conditions.

In contribution 2, Li et al. investigate the effects of different cryoprotectants on the cryopreservation of sperm from the noble scallop *Mimachlamys nobilis*. Six cryoprotectants were tested, including both permeable agents (dimethyl sulfoxide, ethylene glycol, propylene glycol, and methanol) and non-permeable agents (trehalose and fetal bovine serum), at varying concentrations. The results showed that permeable cryoprotectants, especially 10% dimethyl sulfoxide (DMSO), provided the best preservation of sperm motility, velocity, and morphology. Non-permeable cryoprotectants like fetal bovine serum also demonstrated protective effects at higher concentrations. Morphological and ultrastructural analyses revealed that cryopreservation caused varying degrees of damage, with flagellar integrity being crucial for sperm motility. Notably, sperm cryopreserved with 10% DMSO exhibited minimal structural damage compared to the control group, highlighting its effectiveness in maintaining sperm function post-thawing. These findings emphasize the importance of

https://doi.org/10.3390/jmse14111004

optimizing cryoprotectant selection and concentration for the successful cryopreservation of marine species' sperm, thereby contributing to more effective germplasm conservation strategies for aquaculture.

In contribution 3, Li et al. compare the nutritional composition, antioxidant properties, and genetic diversity of three populations of *Penaeus semisulcatus* (green tiger shrimp) from different aquaculture regions in China. The analysis revealed minimal variation in moisture, crude fat, crude protein, and total sugar content among the populations, although ash content exhibited significant variation. The PsHK population demonstrated superior nutritional value, with higher levels of amino acids and fatty acids, especially saturated and monounsaturated fatty acids. In terms of antioxidant capacity, the populations showed no significant differences in enzyme activity. However, PsHK exhibited slightly higher antioxidant enzyme levels, suggesting a stronger ability to cope with oxidative stress. Genetic analysis revealed that PsHK had the highest genetic diversity with more pronounced differentiation from the PsRP population, whose nutritional and genetic advantages are highlighted by these findings, offering valuable insights for future aquaculture and breeding strategies that aim to enhance the quality and sustainability of *P. semisulcatus* production.

In contribution 4, Vasquez et al. investigate the adaptive responses of *Pteria penguin* (winged pearl oyster) to byssal thread trimming, a technique used to stimulate byssal production and shedding. The oysters were divided into small and large size groups and exposed to two conditions: periodic trimming of byssal threads and no trimming. The results show that trimming significantly increased byssal thread production in both size groups, with small oysters producing more threads than larger ones. Trimming also enhanced the byssal stalk diameter, likely due to the merging of new threads with existing stalks. Moreover, trimming induced byssal shedding, particularly in smaller oysters, suggesting that shedding may serve as an adaptive mechanism to recalibrate attachment strength and optimize energy use. These findings provide valuable insights into the physiological and behavioral plasticity of *Pteria penguin*, with potential implications for aquaculture and pearl oyster management.

In contribution 5, Ding et al. present the MCR-UIE, a novel semi-supervised underwater image enhancement method that integrates multimodal contrastive learning (MCL) with a dynamic quality reliability repository (DQR) to optimize the use of labeled and unlabeled data for improving image quality. The MCR-UIE method incorporates several multimodal contrastive losses, leveraging features from perceptual, edge, color, and local texture information, to increase model robustness and avoid overfitting. The dynamic quality reliability repository ensures high-quality pseudo-labels by selecting reliable outputs from the teacher model. Extensive experiments demonstrate that MCR-UIE consistently outperforms existing methods in both full-reference and no-reference metrics across various underwater image datasets, including real-world scenarios like aquaculture environments. The model's superior performance in maintaining natural color reproduction, enhancing image sharpness, and reducing distortion makes it a strong candidate for practical applications, such as intelligent monitoring systems in complex underwater environments. The method's robust performance underscores its potential for real-world deployment in tasks such as fish detection and behavior analysis in aquaculture.

In contribution 6, Liu et al. proposes an improved DeepSORT-based model for multi-target tracking of underwater fish, addressing challenges such as high-density fish, frequent occlusions, and visual similarities between individuals. The model integrates multiple enhancements, including ResNet for feature extraction, Deformable Convolutional Networks v2 (DCNv2) for adaptive receptive fields, Triplet Loss for improved feature discrimination, and Convolutional Block Attention Module (CBAM) to focus on key features. The improved tracking system outperforms traditional methods, with a 4.15% increase in the MOTA metric and an 18.55% improvement in the IDF1 metric. Experimental results demonstrate that the method significantly reduces ID switching and improves tracking stability, especially in complex underwater environments with occlusion and interaction challenges. The model provides a more reliable solution for analyzing fish behavior, with future research focused on further enhancing its performance in extreme occlusion conditions.

In contribution 7, Wu et al. investigate a bio-inspired vertical-axis hydrokinetic turbine, designed using the morphology of a sailfish for aerodynamic optimization. The study applies parametric optimization to analyze four critical design parameters—arcuate angle, installation angle, thickness ratio, and chord-length ratio. The turbine's dual-rotor system integrates centrifugal and axial-flow modules, each optimizing different energy conversion mechanisms: tidal current kinetic energy and wave potential energy. The results reveal that the centrifugal rotor reaches peak efficiency at a tip speed ratio (λ) of 1.25 and Reynolds number (Re) of 1.33×10^5, while the axial-flow rotor achieves optimal efficiency at $\lambda = 1.5$ and $Re = 5.53 \times 10^4$. The study also examines the aerodynamic sensitivity of various parameters, noting that lift is predominantly influenced by the installation angle, while drag is more affected by the arcuate angle. The findings highlight the challenges in balancing aerodynamic efficiency with structural durability, particularly in terms of the influence of angular configurations on flow separation and hydrodynamic forces. The research highlights the potential of sailfish-inspired turbine designs for improving energy extraction in marine environments, providing insights into turbine blade optimization and the trade-offs associated with energy harvesting systems.

In contribution 8, Ma et al. propose a multimodal vision-based monitoring framework for fishery aquaculture cages, aimed at improving perception and segmentation performance under harsh low-visibility underwater conditions. The study develops a novel dual-image dataset that combines Neuromorphic vision (NeuroVI) and RGB images, covering four critical aquaculture scenarios: feeding desire detection, biofouling attachment detection, fence breach detection, and dead fish detection. Building on this dataset, the authors design a dual-image cross-attention network in which NeuroVI images provide spatial attention cues to guide RGB feature extraction, thereby enhancing the network's ability to identify salient regions in visually degraded environments. To further reduce the computational burden introduced by the dual-channel architecture, the study replaces conventional dot-product attention with an element-wise multiplication mechanism, improving efficiency while maintaining strong detection performance. Experimental results show that the proposed framework achieves state-of-the-art performance, including an 8% improvement in mean average precision over existing methods, while also reducing latency to 1.4 ms in the optimized configuration. The ablation and comparative analyses further demonstrate that the integration of SNN-based NeuroVI filtering, cross-attention learning, and lightweight multiplication yields consistent gains in both segmentation accuracy and computational efficiency. Overall, the research highlights the potential of NeuroVI-RGB multimodal fusion for robust aquaculture monitoring, while also acknowledging limitations related to event camera availability, computational overhead, and generalization across diverse underwater environments.

Advances in genetic engineering, artificial intelligence (AI), and marine biotechnology are set to revolutionize deep-sea aquaculture, enhancing its feasibility and sustainability [11–13]. Genetic engineering, and particularly gene editing, can improve species adaptability, making them more resilient to extreme conditions and increasing productivity and survival rates. AI is expected to optimize operations by predicting environmental conditions, monitoring species health, and dynamically adjusting feeding schedules, thereby improving efficiency and reducing waste. Additionally, marine biotechnology promises innovations

https://doi.org/10.3390/jmse14111004

in developing novel feed ingredients and health supplements tailored to deep-sea species, enhancing sustainability and profitability. As these technologies evolve, scaling deep-sea aquaculture to industrial levels becomes increasingly feasible, offering a solution to the growing global demand for seafood while reducing pressure on shallow marine ecosystems. This scalability, along with lower operational costs and improved efficiency, positions deep-sea aquaculture as a key player in global food security and environmental sustainability. The future of deep-sea aquaculture holds great promise. Through strategic species selection, technological innovation, and sustainable practices, this industry can meet the rising demand for seafood while preserving shallow marine ecosystems. Continued research and cross-industry collaboration will be crucial for realizing its full potential, contributing to both food security and environmental sustainability.

Acknowledgments: As Guest Editors of the Special Issue "Selection of Deep-Sea Aquaculture Species and Development of Supporting Technologies and Equipment", we wish to extend our sincere gratitude to all the authors whose valuable contributions made the publication of this issue possible. Their work has significantly enriched and enhanced the coverage of this publication.

Conflicts of Interest: The authors declare no conflicts of interest.

List of Contributions:

1. Huang, J.; Fu, Z.; Yu, W.; Hou, B.; Wu, J.; Zhang, T.; Ma, Z. Gut Microbiota Response to Experimental Acute Cold Stress in Juvenile Yellowfin Tuna (*Thunnus albacares*). *J. Mar. Sci. Eng.* **2025**, *13*, 602. https://doi.org/10.3390/jmse13030602.
2. Li, M.; Fu, Z.; Yu, G.; Ma, Z. Effects of Different Cryoprotectants on Cryopreservation of Sperm from Noble Scallop *Mimachlamys nobilis*. *J. Mar. Sci. Eng.* **2025**, *13*, 607. https://doi.org/10.3390/jmse13030607.
3. Li, Y.; Chen, J.; Jiang, Z.; Jiang, S.; Yang, Q.; Yang, L.; Huang, J.; Shi, J.; Ding, Y.; Liu, T.; et al. Comparative Analysis of Amino Acid Composition, Fatty Acid Profiles, and Genetic Diversity Among Three Populations of *Penaeus semisulcatus*. *J. Mar. Sci. Eng.* **2025**, *13*, 655. https://doi.org/10.3390/jmse13040655.
4. Vasquez, H.E.; Wei, S.; Yang, G.; Wang, L.; Yu, P.; Dong, M.; Yuan, C.; Zheng, X. Adaptive Responses in Byssal Growth and Shedding: Insights from *Pteria penguin* Under Thread Trimming and Non-Trimming Conditions. *J. Mar. Sci. Eng.* **2025**, *13*, 874. https://doi.org/10.3390/jmse13050874.
5. Ding, M.; Li, G.; Hu, Y.; Liu, H.; Hu, Q.; Huang, X. Semi-Supervised Underwater Image Enhancement Method Using Multimodal Features and Dynamic Quality Repository. *J. Mar. Sci. Eng.* **2025**, *13*, 1195. https://doi.org/10.3390/jmse13061195.
6. Liu, S.; Zhang, J.; Zheng, H.; Qian, C.; Liu, S. An Improved DeepSORT-Based Model for Multi-Target Tracking of Underwater Fish. *J. Mar. Sci. Eng.* **2025**, *13*, 1256. https://doi.org/10.3390/jmse13071256.
7. Wu, A.; Wang, S.; Ding, C. The Hydrodynamic Performance of a Vertical-Axis Hydro Turbine with an Airfoil Designed Based on the Outline of a Sailfish. *J. Mar. Sci. Eng.* **2025**, *13*, 1266. https://doi.org/10.3390/jmse13071266.
8. Ma, F.; Liu, X.; Xu, Z. A Multimodal Vision-Based FishEnvironment and Growth Monitoring in an Aquaculture Cage. *J. Mar. Sci. Eng.* **2025**, *13*, 1700. https://doi.org/10.3390/jmse13091700.

References

1. Gentry, R.R.; Froehlich, H.E.; Grimm, D.; Kareiva, P.; Parke, M.; Rust, M.; Gaines, S.D.; Halpern, B.S. Mapping the global potential for marine aquaculture. *Nat. Ecol. Evol.* **2017**, *1*, 1317–1324. [CrossRef] [PubMed]
2. Naylor, R.L.; Hardy, R.W.; Buschmann, A.H.; Bush, S.R.; Cao, L.; Klinger, D.H.; Little, D.C.; Lubchenco, J.; Sumway, S.E.; Troell, M. A 20-year retrospective review of global aquaculture. *Nature* **2021**, *591*, 551–563. [CrossRef] [PubMed]
3. Qin, J.; Ma, Z.; Li, Y. Review and Outlook for Research and Development in Aquatic Life and Ecosystems. *ALE* **2025**, *1*, 2. [CrossRef]

https://doi.org/10.3390/jmse14111004

4. Nichols, C.R.; Zinnert, J.; Young, D.R. Degradation of Coastal Ecosystems: Causes, Impacts and Mitigation Efforts. In *Tomorrow's Coasts: Complex and Impermanent*; Wright, L., Nichols, C., Eds.; Coastal Research Library; Springer: Cham, Switzerland, 2019; Volume 27. [CrossRef]
5. Plessis, A. Persistent degradation: Global water quality challenges and required actions. *One Earth* **2022**, *5*, 129–131. [CrossRef]
6. Sumaila, U.R.; Tai, T.C. End Overfishing and Increase the Resilience of the Ocean to Climate Change. *Front. Mar. Sci.* **2020**, *7*, 523. [CrossRef]
7. Flies, E.J.; Brook, B.W.; Blomqvist, L.; Buettel, J.C. Forecasting future global food demand: A systematic review and meta-analysis of model complexity. *Environ. Int.* **2018**, *120*, 93–103. [CrossRef] [PubMed]
8. Yu, J.; Yan, T. Analyzing Industrialization of Deep-Sea Cage Mariculture in China: Review and Performance. *Rev. Fish. Sci. Aquac.* **2023**, *31*, 483–496. [CrossRef]
9. Fujita, R.; Brittingham, P.; Cao, L.; Froehlich, H.; Thompson, M.; Voorhees, T. Toward an environmentally responsible offshore aquaculture industry in the United States: Ecological risks, remedies, and knowledge gaps. *Mar. Pol.* **2023**, *147*, 105351. [CrossRef]
10. Wang, C.M.; Chu, Y.; Nguyen, H.P. Advancements in Offshore Aquaculture: Enhancing Environmental Sustainability and Elevating Seafood Production. In *Proceedings of the Fourth World Conference on Floating Solutions: WCFS 2024, 2–4 December, Hong Kong, WCFS 2024*; Zhao, X.L., Dai, J.G., Liu, S., Lim, S.H., Eds.; Lecture Notes in Civil Engineering; Springer: Singapore, 2025; Volume 597. [CrossRef]
11. Ditria, E.M.; Buelow, C.A.; Gonzalez-Rivero, M.; Connolly, R.M. Artificial intelligence and automated monitoring for assisting conservation of marine ecosystems: A perspective. *Front. Mar. Sci.* **2022**, *9*, 918104. [CrossRef]
12. Holzinger, A.; Keiblinger, K.; Holub, P.; Zatloukal, K.; Müller, H. AI for life: Trends in artificial intelligence for biotechnology. *New Biotechnol.* **2023**, *74*, 16–24. [CrossRef] [PubMed]
13. Grof-Vindman, C.S.; Trump, B.D.; Cummings, C.L.; Smith, M.; Titus, A.J.; Oye, K.; Prado, V.; Turmus, E.; Linkov, I. The convergence of AI and synthetic biology: The looming deluge. *npj Biomed. Innov.* **2025**, *2*, 20. [CrossRef] [PubMed]

https://doi.org/10.3390/jmse14111004

Journal of
*Marine Science
and Engineering*

MDPI

Article

Gut Microbiota Response to Experimental Acute Cold Stress in Juvenile Yellowfin Tuna (*Thunnus albacares*)

Junhua Huang [1,3,4,5,6], Zhengyi Fu [1,3,4,5,7], Wei Yu [1,3,4,5], Bowen Hou [8], Jinhui Wu [2], Tao Zhang [9,*] and Zhenhua Ma [1,3,4,5,7,*]

1 Key Laboratory of Efficient Utilization and Processing of Marine Fishery Resources of Hainan Province, Sanya Tropical Fisheries Research Institute, Sanya 572018, China; junhua_huang@sina.cn (J.H.); fu0174@flinders.edu.au (Z.F.); weiyu88888888@126.com (W.Y.)
2 Agro-Tech Extension Center of Guangdong Province, Guangzhou 510000, China; wjhin@sina.com
3 South China Sea Fisheries Research Institute, Chinese Academy of Fishery Sciences, Guangzhou 510300, China
4 Hainan Engineering Research Center for Deep-Sea Aquaculture and Processing, Sanya Tropical Fisheries Research Institute, Sanya 572018, China
5 International Joint Research Center for Conservation and Application of Fishery Resources in the South China Sea, Sanya Tropical Fisheries Research Institute, Sanya 572018, China
6 College of Fisheries and Life Sciences, Dalian Ocean University, Dalian 116023, China
7 College of Science and Engineering, Flinders University, Adelaide 5001, Australia
8 Hainan Haiken Marine Industry Group Co., Ltd., Haikou 570000, China; houbw97@163.com
9 Dalian Tianzheng Industrial Co., Ltd., Dalian 116021, China
* Correspondence: zht_3000@163.com (T.Z.); zhenhua.ma@scsfri.ac.cn (Z.M.)

Abstract: Environmental temperature declines significantly impact the physiological processes of marine organisms, particularly under the cool La Nina conditions, challenging yellowfin tuna (*Thunnus albacares*) aquaculture. Low temperatures affect fish metabolism and immune functions and can alter the gut microbiota composition, influencing health and growth. This study investigates the impact of low temperatures on the gut microbiota of juvenile yellowfin tuna. Fish were divided into a Low Temperature (LT) group (24 °C), an Ultra Low Temperature (ULT) group (18 °C), and a Control group (CG) (30 °C), with evaluations at 0, 12, 24, and 36 h using α-diversity analysis and microbial species composition. Results indicated a significant increase in the Ace index for the ULT group after 36 h of cold stress ($p < 0.05$), with no significant changes in the Shannon index. A decline in Proteobacteria and increases in Verrucomicrobiota and Firmicutes were observed in both LT and ULT groups. Additionally, both LT and ULT groups showed a significant rise in the Ace index at 36 h ($p < 0.05$), with a significant decrease in the Shannon index in the ULT group at 24 h. Furthermore, Firmicutes significantly increased at 12 h in both temperature groups ($p < 0.05$). These findings highlight the potential role of the gut microbiome in adapting yellowfin tuna to cold environments and provide microbial insights into their physiological adaptations, laying a foundation for further research and practical applications in aquaculture under cold conditions.

Keywords: yellowfin tuna; low-temperature stress; gut microbiota; alpha diversity; species composition

1. Introduction

The microbial communities within fish are diverse, comprising protozoa, fungi, yeasts, viruses, bacteria, and archaea [1]. Among these, bacteria are the most abundant microbes in the fish gut, playing crucial roles in digestion, immune regulation, and nutrient metabolism [2]. Compared to terrestrial animals, the gut microbial communities in fish exhibit greater complexity across different aquatic environments. Aquatic microbes can interact with the host through fish gills, skin, and intestines, forming a dynamically balanced ecosystem [3]. Studies have indicated that gut microbes are involved in physiological processes such as vitamin synthesis, lipid metabolism, and nitrogen cycling within the fish [4]. However, tuna possess a unique thermoregulatory mechanism known as regional endothermy, which allows them to maintain higher temperatures in certain tissues, such as the brain and muscles, compared to the surrounding environment [5]. This thermoregulatory ability is not present in all tuna species but is particularly evident in those within the Thunnini tribe. This adaptation enables tuna to function efficiently in colder waters, avoiding thermal stress that typically affects ectothermic fish [6]. The thermoregulatory mechanism in tuna may influence the response of their gut microbiota to cold stress, as warmer body tissues experience less temperature fluctuation. In contrast, fish that do not regulate their body temperature, such as *Cyprinus carpio* and *Micropterus salmoides*, may undergo more significant changes in microbiota composition and activity under cold stress [7]. Nonetheless, current research on the impacts of cold stress on the gut microbiota of tuna remains insufficient. Most studies on the gut microbiota and temperature stress in fish have focused on other ectothermic species [8]. Research has shown that in ectothermic species like zebrafish (*Danio rerio*), the composition and function of the gut microbiota change under low-temperature conditions, particularly the abundance of microbes related to immune regulation and metabolism [9,10]. However, similar research is scarce for species with advanced thermoregulatory abilities, such as tuna.

Yellowfin tuna (*Thunnus albacares*), a high-speed pelagic species, not only holds a significant position in the global fishery market but also serves as an important apex predator in marine ecosystems [11]. Widely distributed across global tropical and subtropical waters, yellowfin tuna thrive in warm waters and exhibit strong temperature adaptability [12]. Studies have shown that the optimal growth temperature for yellowfin tuna in the Western Indian Ocean typically ranges from 20 to 30 °C. Within this temperature range, their metabolic rate remains high, their growth rate is maximized, and their gut microbiota maintains greater stability and diversity. However, temperatures below 18 °C may negatively impact metabolism, growth, and the stability of the gut microbial community [13]. However, due to the variability of marine environments, their habitat is often affected by external climate changes, particularly temperature fluctuations [14]. Influenced by the La Nina phenomenon, significant low-temperature variations have recently occurred in the South China Sea [15]. La Nina typically leads to cooler sea temperatures in the central and eastern equatorial Pacific, triggering global climate anomalies, especially intensifying cold air activities in Asia [16]. These abnormal low-temperature variations disrupt the marine ecosystem in the South China Sea and profoundly affect aquatic life. Water temperature is a critical ecological factor that influences the growth, reproduction, and survival of marine fish, and drastic changes can impose significant physiological and metabolic stress, forcing fish to adapt to these environmental shifts [17,18]. For fish, low temperatures pose not just a physical challenge but also a biological stress, significantly impacting their metabolic rate, growth, reproduction, and behavioral patterns [19]. In cold environments, the gut microbial communities in fish are also significantly affected [20]. Research indicates that low temperatures can alter the structure of microbial communities within the fish gut, affecting the diversity and functionality of these microbes [8]. In a study by Lv et al. on large yellow croaker (*Larimichthys crocea*), when water temperatures dropped from 12 °C to 8 °C, significant changes occurred in the diversity and abundance of gut microbiota [21]. These findings further demonstrate the impact of low temperatures on fish health and adaptability through changes in gut microbial communities.

Under cold stress conditions, metabolic activities in fish may slow down, altering the flow of nutrients, and thereby affecting the balance of microbial communities [22]. Additionally, microbes with cold tolerance may become selectively enriched in the gut under certain low-temperature conditions, while species coexisting in warmer waters might diminish or disappear [23]. These microbial changes can affect not only the efficiency of digestion and absorption in fish but also impact their immune systems, as gut microbes play a crucial role in regulating host immune responses [24,25]. Studies have shown that under low temperatures, a decrease in beneficial bacteria and an increase in harmful bacteria can lead to dysbiosis, further triggering intestinal health issues such as impaired gut barrier functions and heightened inflammatory responses [26,27]. For juvenile yellowfin tuna, cold stress could exacerbate physiological burdens such as impaired metabolism, reduced immune function, and compromised cellular processes, further affecting the stability and functionality of their gut microbiota. Moreover, the gut microbiota of juveniles is not fully matured, making them relatively vulnerable to environmental changes and susceptible to the effects of cold stress [28].

In this study, we set the minimum experimental water temperature at 18 °C. This decision is based on actual observations of winter water temperatures in the northern South China Sea under the influence of extreme cold currents. Under La Nina conditions, sea surface temperatures in this region can drop to 18 °C, representing one of the lowest temperatures that yellowfin tuna may encounter in their natural environment [29]. Despite this, there remains limited research on the impact of low-temperature stress on the gut microbiota of yellowfin tuna. Therefore, our study compared the effects on gut microbiota at two different low temperatures (24 °C and 18 °C) relative to normal water temperature (30 °C) over 36 h, as well as the distribution characteristics of gut microbiota communities at 0, 12, 24, and 36 h under the two different low-temperature conditions. Utilizing high-throughput sequencing technology to analyze the gut microbiota, we focused on α-diversity and community distribution characteristics under different temperature treatments to clarify the impact of low-temperature stress on microbial community structure and function. Additionally, we explored the abundance changes in key functional microbial groups involved in metabolism and immune regulation under low-temperature stress conditions. Through these analyses, this study provides new insights into the physiological adaptation mechanisms of yellowfin tuna in response to climate change, offers foundational information on changes in the gut microbiota of yellowfin tuna in cold environments, and further supports the optimization of aquaculture environments and management strategies for yellowfin tuna.

2. Materials and Methods

2.1. Experimental Fish and Design

For this experiment, 180 juvenile yellowfin tuna (average body length: 30.26 ± 2.79 cm; average weight: 384.88 ± 58.04 g) were wild-caught from the waters off Xincun Town, Lingshui Li Autonomous County, Hainan Province. The fish were then provided by the Tropical Aquatic Product Research and Development Center of the South China Sea Fisheries Research Institute, China. The juvenile yellowfin tuna were placed in circular recirculating aquaculture tanks with a radius of 1.5 m and a height of 1.5 m, filled with 10 m^3 of filtered seawater, and maintained with a daily water exchange rate of 300%. Before the experiment, the fish underwent a 7-day acclimation period in the experimental tanks. During acclimation, they were fed fresh chopped fish (2 × 2 cm) once daily at 8:00 a.m., at a feeding rate of 5–8% of body weight to satiation. Feeding was suspended one day before the experiment and throughout the trial. During the acclimation period, the fish were closely monitored, with no mortality or abnormal stress behaviors observed, and their feeding activity remained stable.

The experiment was divided into three groups: CG (30 °C, ambient water temperature), LT (24 °C, low temperature), and ULT (18 °C, ultra-low temperature). Each group had three replicate tanks, with 15 juvenile yellowfin tuna each, and lasted for 36 h. During the experiment, salinity was maintained at 32‰, dissolved oxygen (DO) > 7.50 mg·L^{-1}, pH at 7.93 ± 0.12, ammonia nitrogen < 0.1 mg·L^{-1}, and nitrite nitrogen < 0.05 mg·L^{-1}. Temperature reduction in LT and ULT groups was achieved using a chiller (Haier, Qingdao, China) at a rate of 2 °C per hour until the set temperatures were reached. Sampling occurred at 0, 12, 24, and 36 h during cold exposure, with three fish randomly selected from each tank for analysis. The selected fish were anesthetized with an overdose of MS-222, and their intestines were rapidly excised for microbiota analysis. The excised intestinal tissues were immediately rinsed with sterile water, and their contents were carefully transferred into 1.5 mL sterile centrifuge tubes, which were then stored at −80 °C until further analysis.

2.2. DNA Extraction, PCR Amplification, and Construction of a 16S rDNA Library

According to the protocol provided with the DNA kit, total DNA was extracted from the intestinal contents of 36 samples using the TIANGEN Marine Animal Tissue DNA Extraction Kit (Tiangen Biotech Co., Beijing, China). The quantity and quality of the DNA were assessed using the Qubit 3.0 Fluorometer and agarose gel electrophoresis.

Specific primers designed for the Illumina MiSeq platform were used to PCR amplify the bacterial 16S rDNA gene V3-V4 hypervariable regions (Forward primer: 5′-ACTCCTACGGGAGGCAGCAG-3′, Reverse primer: 5′-GGACTACHVGGGTWTCTAAT-3′). Sequencing adapters were added to the PCR products to facilitate subsequent MiSeq sequencing. All PCR amplifications were performed in a 25 μL reaction mixture containing: 2.5 μL TransStart buffer, 2 μL dNTP mix, 1 μL of each primer, and 20 ng template DNA. The thermocycling conditions were as follows: initial denaturation at 94 °C for 3 min, denaturation at 94 °C for 5 min, annealing at 57 °C for 90 min, extension at 72 °C for 10 s, and a final extension at 72 °C for 5 min. The PCR amplicons were assessed for quality using 1.5% (*w*/*v*) agarose gel electrophoresis and purified with the Quick Gel Extraction Kit (Qiagen, Hilden, Germany). To ensure the reliability of sequencing data, a series of quality control measures were conducted prior to sequencing. DNA concentration and purity were measured using a Qubit 3.0 fluorometer, and DNA integrity was evaluated via agarose gel electrophoresis. Only samples meeting the following criteria were selected for downstream library construction and sequencing: DNA concentration ≥ 10 ng/μL, A260/A280 ratio between 1.8 and 2.0, and intact DNA bands without significant degradation. Samples that did not meet these criteria were excluded from further analysis. Additionally, no-template controls (NTCs) were included in the PCR process to detect potential contamination. Only PCR products that met the required concentration and quality standards were used for library construction and high-throughput sequencing.

2.3. Bioinformatics Analysis

The gene library concentration was measured using the Qubit 3.0 spectrophotometer and normalized to 10 nM. According to the instructions, it was then loaded onto an Illumina MiSeq system for high-throughput sequencing (Illumina, San Diego, CA, USA), using PE 300 for paired-end sequencing. Image analysis and base calling were conducted using the MiSeq Control Software (version 4.1.0) included with the MiSeq system [30]. Post-sequencing, paired-end read frames were added, and samples were classified based on barcodes, trimming barcodes, and primer sequences [31]. Quality filtering was performed based on linker sequences, removing reads with ambiguous bases, a quality score of Q ≥ 20, or shorter than 200 bp. Raw data were merged using Flash (version v1.2.7) and filtered with Qiime (version v1.9.1). Uchime analysis was then conducted to remove chimeric clusters from the sequencing data of each sample [32]. Effective data with 99% sequence identity were clustered into operational taxonomic units (OTUs) using Uparse (version 7.0.1001), and OTU taxonomic assignment was completed using the Ribosomal Database Project (RDP) classifier [33]. Representative sequences of OTUs were aligned using the Silva_128 16S rRNA database [34]. Rarefaction curves were analyzed with mothur (version v.1.30). Bacterial α-diversity indices, including the Shannon index (diversity), abundance-based coverage estimator (Ace index, richness), and coverage (Good's coverage), were calculated using Qiime. Effective data were clustered into OTUs at 99% similarity using Uparse, and species annotation was performed using the RDP Classifier tool, based on the Green Genes database. Beta diversity analyses were conducted to compare microbial communities among different samples. Heatmaps were generated using the 'ggplot2' package in R [35], and principal component analysis was performed using UniFrac principal component analysis. The version of R used was R 4.0.5.

2.4. Statistical Analysis

Data were organized using Microsoft Excel 2021 and analyzed using SPSS version 26.0. All values are presented as means ± standard deviation (mean ± SD). For microbial community analysis, statistical differences were assessed using PERMANOVA with $p < 0.05$ considered statistically significant. The plots were generated using Origin 2021.

3. Results

3.1. Distribution of the 16S rRNA Gene Library

Following the processing of the 36 samples, a total of 2,126,867 effective sequences were obtained, with the number of sequences per sample ranging from 45,983 to 152,254, and an average sequence length of 421.13 bp (Table S1). Rarefaction curves (Figure 1) approached an asymptote as sequencing depth increased, indicating that additional sequencing would yield minimal new operational taxonomic units (OTUs), thus confirming sequencing saturation. Additionally, Good's coverage values for all samples exceeded 99.8% (Table S2), further supporting that sequencing depth was sufficient to comprehensively capture microbiota diversity.

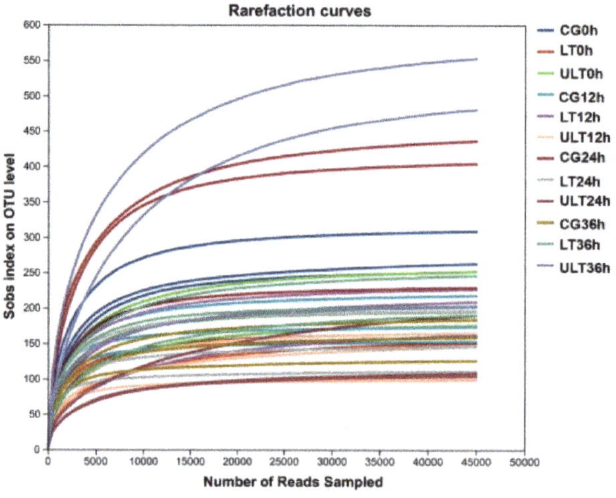

Figure 1. Rarefaction curves of gut content samples from the CG, HT group, and ULT group under different durations of cold stress.

3.2. Analysis of α-Diversity and the Number of OTUs Among Different Treatment Groups

At 36 h into the experiment, α-diversity analyses were conducted for different experimental treatment groups (Figure 2). The Ace and Shannon indices, reflecting community richness and both diversity and evenness, respectively, were analyzed for the CG, LT group, and ULT group at 36 h of cold stress. Based on the Ace index (Figure 2A), the ULT group exhibited a significantly higher Ace index compared to the CG ($p < 0.05$), while the differences between CG and LT groups as well as between LT and ULT groups were not significant ($p > 0.05$). In the analysis based on the Shannon index (Figure 2B), there were no significant differences in the Shannon index among the three groups ($p > 0.05$).

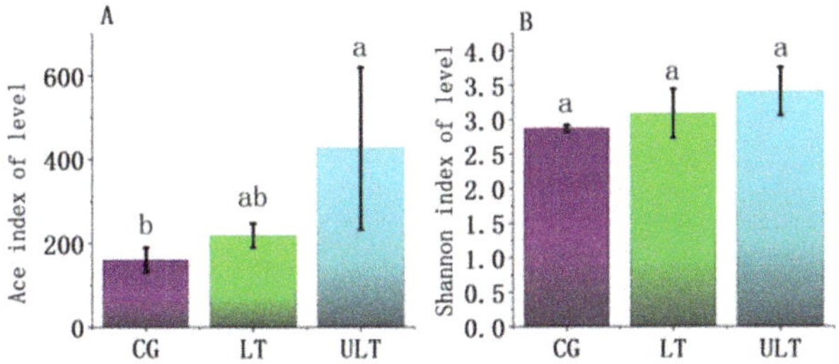

Figure 2. Analysis of α-diversity among different treatment groups ((**A**) Ace index diversity analysis; (**B**) Shannon index diversity analysis; different lowercase letters indicate significant differences between groups, $p < 0.05$).

A Venn diagram was constructed to analyze the distribution of OTUs among the CG, HT group, and ULT group at 36 h (Figure 3). The total number of OTUs for the CG, HT group, and ULT group were 317, 389, and 932, respectively, consistent with the Ace and Shannon index results. After 36 h, there were 25 common OTUs between the CG and LT group, 46 common OTUs between the CG and ULT group, and 71 common OTUs between the LT and ULT groups. Specifically, there were 120 unique OTUs in the CG, 167 in the LT group, and 689 in the ULT group, with 126 OTUs shared among all three groups.

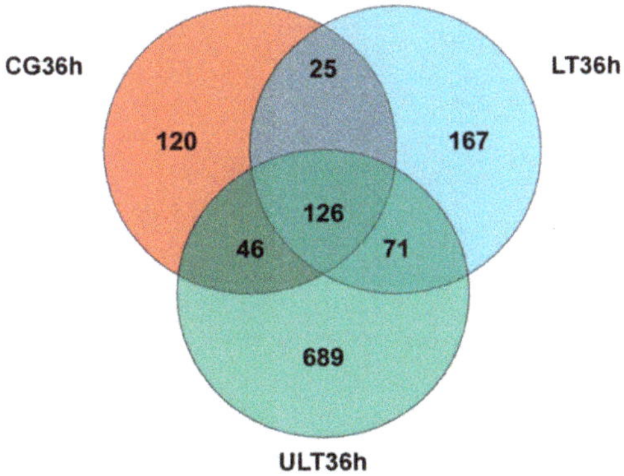

Figure 3. Venn diagram analyzing the number of OTUs at 36 h among the CG, HT group, and ULT group.

3.3. Diversity and Composition of Gut Microbiota Among Different Treatment Groups

After 36 h of cold stress, the gut microbiota compositions of the different experimental groups varied at the phylum level. As shown in Figure 4, *Proteobacteria* was the most predominant phylum in the CG, LT group, and ULT group after 36 h of cold stress. With decreasing temperature, the proportion of *Proteobacteria* decreased, accounting for 93.6%,

83.7%, and 81.7% of the total bacterial count in the CG, HT group, and ULT group, respectively. Proteobacteria, Verrucomicrobiota, and Bacteroidota were dominant phyla in both the CG and ULT group. In the CG, Verrucomicrobiota and Bacteroidota represented 1.98% and 1.84% of the total bacterial count, respectively, and 12.53% and 1.61% in the LT group. However, there was a shift in the dominant phyla in the ULT group, where Proteobacteria, Firmicutes, and Bacteroidota were prevalent. In the ULT group, Firmicutes and Bacteroidota accounted for 9.05% and 3.73% of the total bacterial count, respectively.

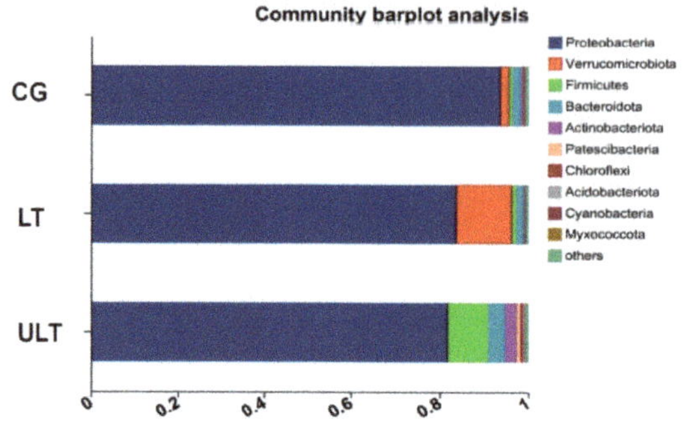

Figure 4. Bar graph analysis of gut microbiota composition at the phylum level among different treatment groups.

Cluster analysis of the gut microbiota at the genus level was conducted for the three treatment groups (Figure 5), including the 30 most abundant microbial genera in each group. The analysis revealed that the dominant genera in the CG were primarily Acinetobacter, Brevundimonas, Pseudomonas, and Sphingomonas. In the LT group and ULT group, the dominant genera differed from those in the CG. In the LT group, the predominant genera were *Acinetobacter*, *Pseudomonas*, and *Trichococcus*. In the ULT group, the dominant genera were *Acinetobacter*, *Persicirhabdus*, *Pseudomonas*, and *Sphingomonas*.

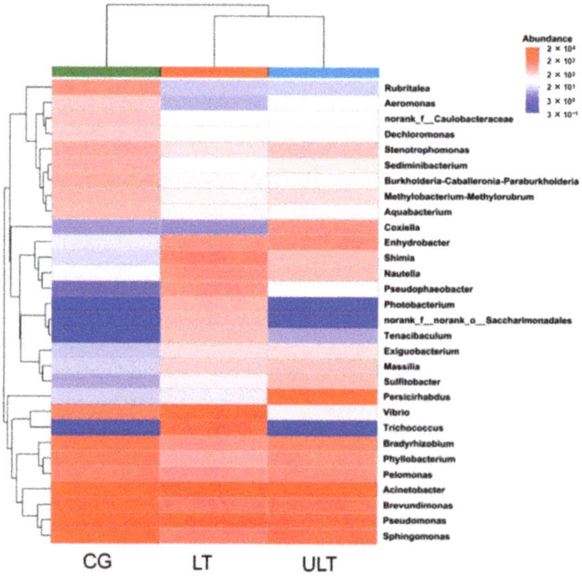

Figure 5. Heatmap of the abundance of gut microbiota at the genus level across different treatment groups.

3.4. Analysis of α-Diversity and Distribution of OTUs over Time in Different Low-Temperature Treatment Groups

Analysis of α-diversity was conducted for two different low-temperature groups at various time intervals (Figure 6). After 36 h of cold stress, the LT group showed a significantly lower Ace index at 0 h compared to 12 h and 36 h ($p < 0.05$) (Figure 6A). The Shannon index in the LT group did not show significant differences across the stress periods ($p > 0.05$) (Figure 6B). For the ULT group, as depicted in Figure 7, the Ace index was significantly higher at 36 h compared to 12 h and 24 h ($p < 0.05$), but there was no significant difference compared to 0 h ($p > 0.05$). The Shannon index in the ULT group was significantly lower at 24 h compared to other time intervals ($p < 0.05$) (Figure 6D).

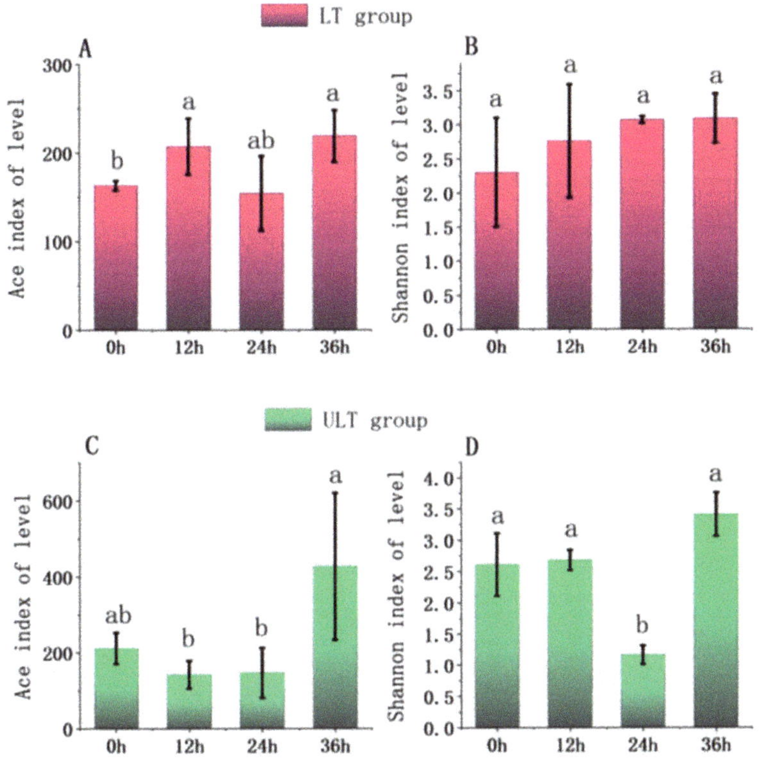

Figure 6. Changes in richness and diversity indices of gut microbiota over time in LT and ULT groups ((**A**) Ace index for LT group; (**B**) Shannon index for LT group; (**C**) Ace index for ULT group; (**D**) Shannon index for ULT group; different lowercase letters indicate significant differences between groups, $p < 0.05$).

Venn diagrams analyzing the distribution of OTUs over time under different low-temperature conditions are shown in Figure 7. In the LT group, the numbers of OTUs at 0, 12, 24, and 36 h were 320, 401, 305, and 389, respectively, with unique OTUs at these time points numbering 132, 157, 105, and 139, and 86 OTUs common across all four time points (Figure 7A). In the ULT group, the numbers of OTUs at 0, 12, 24, and 36 h were 420, 265, 262, and 932, respectively, with unique OTUs at these intervals numbering 179, 84, 67, and 656. The ULT group had 88 OTUs common across all four different time intervals (Figure 7B).

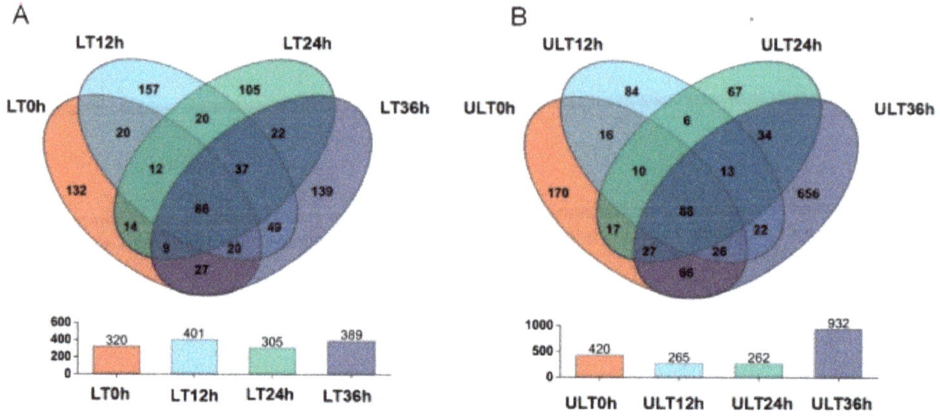

Figure 7. Venn diagram analysis of OTU numbers at different time intervals under varying low-temperature stress conditions, with bar graphs showing the total number of OTUs at each time interval ((**A**) LT group; (**B**) ULT group).

3.5. Diversity and Analysis of Gut Microbiota in Different Low-Temperature Treatment Groups

In the LT group's phylum-level microbial community analysis, *Proteobacteria* was the dominant phylum across all time points, with an increase in the proportion of *Firmicutes* at 12 h, accounting for 24.5% of the total microbial population at that time before returning to baseline levels at 24 and 36 h. Verrucomicrobiota increased to 12.5% of the total microbial population at 36 h (Figure 8A). As shown in Figure 8B, significant differences were observed only in Verrucomicrobiota across different time intervals ($p < 0.05$). At the genus level, *Vibrio* showed a declining trend over time under cold stress, whereas *Mycoplasma* significantly increased at 12 h ($p < 0.05$), becoming the dominant genus at this time point before returning to initial levels at 24 and 36 h. *Acinetobacter* showed a rising trend at 24 and 36 h, becoming the dominant genus at these time points (Figure 8C). According to Figure 8D, significant differences were observed in *Mycoplasma*, *Persicirhabdus*, *Enhydrobacter*, *Photobacterium*, and *Coxiella* across different time intervals ($p < 0.05$).

In the ULT group's phylum-level microbial community analysis, *Proteobacteria* remained the dominant phylum across all time points. Notably, there was an increase in the proportion of *Firmicutes* at 12 and 36 h, accounting for 14.4% and 9.1% of the total bacterial count, respectively. Additionally, the proportions of Bacteroidota and Actinobacteriota also increased at 36 h (Figure 9A). As shown in Figure 9B, significant differences were observed in Actinobacteriota, Spirochaetota, and Chloroflexi across different time intervals ($p < 0.05$). At the genus level, *Vibrio* was the dominant genus at 0, 12, and 24 h, reaching its highest proportion at 24 h, comprising 76.4% of the total bacterial count. However, by 36 h, *Acinetobacter* emerged as the dominant genus, significantly increasing at this time point ($p < 0.05$) and accounting for 30% of the total bacterial count. Additionally, *Pseudomonas* showed a declining trend at 0, 12, and 24 h, returning to its original level by 36 h (Figure 9C). According to Figure 9D, significant differences in the relative abundances of *Acinetobacter*, *Sphingomonas*, *Brevundimonas*, *Bradyrhizobium*, *Photobacterium*, *Enterococcus*, and *Phyllobacterium* were observed across different time intervals within the ULT group ($p < 0.05$).

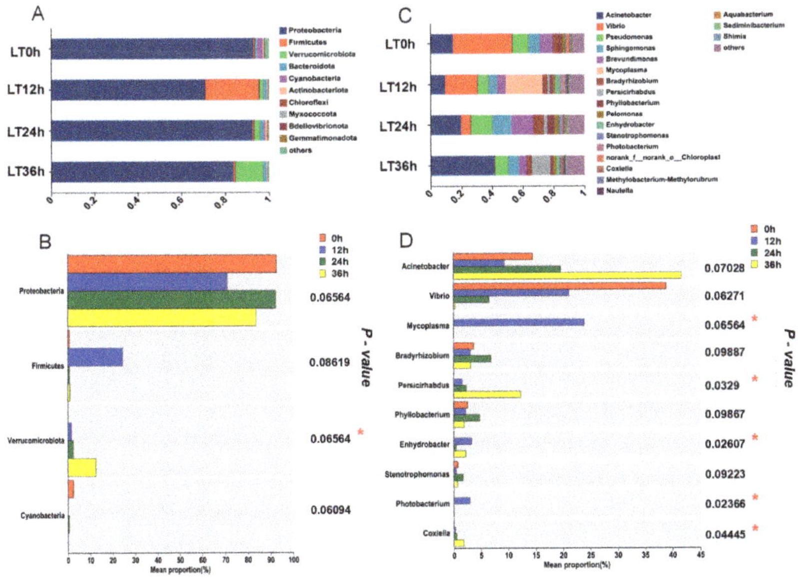

Figure 8. Temporal changes in the gut microbiota composition of the LT group at phylum and genus levels ((**A**) Bar graph of gut microbiota composition at the phylum level in the LT group; (**B**) differences in average relative abundance at the phylum level in the LT group; (**C**) bar graph of gut microbiota composition at the genus level in the LT group; (**D**) differences in average relative abundance at the genus level in the LT group; '*' indicates significant differences, $p < 0.05$).

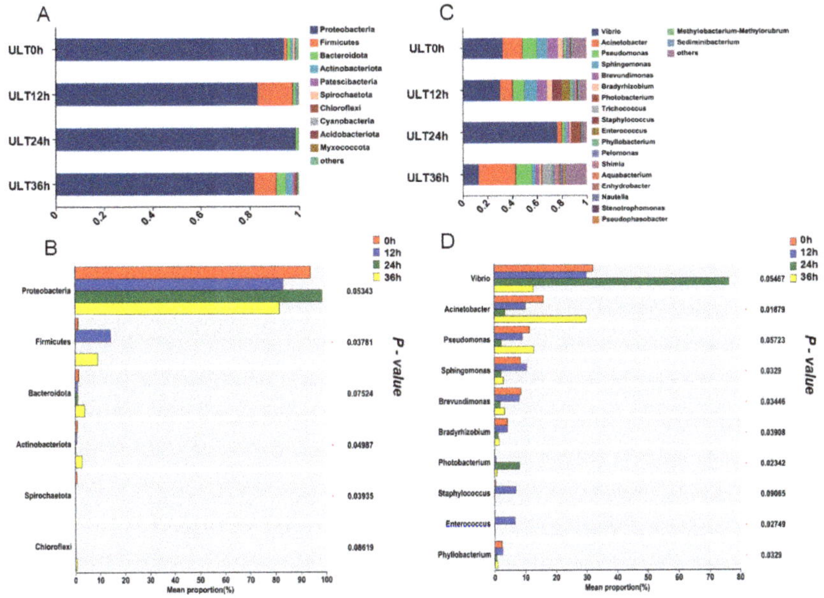

Figure 9. Temporal changes in the gut microbiota composition of the ULT group at phylum and genus levels ((**A**) Bar graph of gut microbiota composition at the phylum level in the ULT group; (**B**) Differences in average relative abundance at the phylum level in the ULT group; (**C**) Bar graph of gut microbiota composition at the genus level in the ULT group; (**D**) Differences in average relative abundance at the genus level in the ULT group; '*' indicates significant differences, $p < 0.05$).

4. Discussion

The gut microbiota of animals comprises a complex and diverse array of microbial communities that form a crucial functional unit, playing significant roles in host growth, metabolism, and immunity [36,37]. Under cold environmental conditions, the composition and functionality of the gut microbiota undergo significant changes, often closely related to the host's physiological stress responses [38]. Previous studies have shown that in golden pompano (*Trachinotus ovatus*), the composition of the main microbial communities in the gut changes under low-temperature stress at 18 °C, although the abundance of these communities does not show significant changes [39]. In contrast, research on yellow catfish (*Pelteobagrus fulvidraco*) has revealed a decline in gut microbiota abundance under cold stress, with Plesiomonas dominating the microbial community in such conditions [40]. Additionally, Ma et al. found that low-temperature stress led to a significant decrease in the relative abundance of Firmicutes and Verrucomicrobiota, while increasing the relative abundance of Clostridia, Actinobacteria, and Proteobacteria in the gut microbiota of large-mouth bass (*Micropterus salmoides*) [41]. Additionally, under high-temperature conditions, Koji Murashita et al. found that trypsin, chymotrypsin, and amylase activity decreased in juvenile Pacific bluefin tuna (*Thunnus orientalis*) [42]. In our study, after 36 h of acute cold stress, the composition and abundance of the gut microbiota in juvenile yellowfin tuna exhibited significant changes. The Ace index, a key indicator of microbial community richness, typically increases when low-abundance species become more prevalent [43]. Under cold stress conditions, alterations in host physiological status—such as changes in intestinal mucus secretion, digestive enzyme activity, and immune responses—can modulate the gut environment, thereby influencing the microbial community structure [44]. However, contrary to the expected trend, the Ace index in the ULT group increased significantly at 36 h. This suggests that certain low-abundance or rare bacterial taxa may have expanded under cold stress, or that shifts in microbial community dynamics contributed to an overall increase in diversity. Moreover, Proteobacteria remained the dominant phylum across all groups, yet its relative abundance decreased under cold conditions. In contrast, the proportions of Firmicutes and Bacteroidota increased significantly in the ULT group. Previous studies have suggested that *Proteobacteria*, as a metabolically active bacterial group, can efficiently utilize nutrients in the intestinal environment through rapid metabolic responses, particularly in warm conditions [45]. The observed decline in Proteobacteria under cold stress may indicate a temperature-dependent shift in microbial metabolism, favoring bacterial taxa that are more adapted to lower temperatures. These findings highlight the complex interplay between environmental stressors and microbial community dynamics, warranting further investigation into the functional implications of these compositional shifts [46,47]. Additionally, under cold conditions, the overall metabolic rate of fish decreases, and the levels of oxygen and nutrient supply in the intestines might reduce, thereby limiting the growth and reproduction of Proteobacteria [48]. This dependency on high metabolic demands makes them less adaptable under cold conditions [49]. Moreover, changes in the gut microenvironment under cold conditions, such as changes in pH and redox potential, can affect the survival advantage of Proteobacteria [50]. Conversely, microbial groups like Firmicutes and Bacteroidota, with their strong metabolic diversity, can efficiently utilize complex carbohydrates and other metabolic substrates through fermentation pathways in energy-limited environments, giving them a competitive advantage under low temperatures [51,52]. Studies have shown that Firmicutes possess a strong ability to degrade carbohydrates and produce short-chain fatty acids (SCFAs), which serve as an energy source for the intestinal epithelium while also regulating host energy metabolism and immune function. Meanwhile, Bacteroidota play a crucial role in maintaining gut microbial balance, degrading polysaccharides, and modulating host immune responses [53]. The significant increase in Firmicutes and Bacteroidota may represent a cold adaptation strategy, enhancing the gut microbiota's ability to break down complex carbohydrates, maintain intestinal barrier stability, and regulate host immune function, thereby optimizing the physiological state of fish under cold stress. Additionally, at the genus level, *Acinetobacter* dominates under cold conditions. Bacteria of the genus *Acinetobacter* are known for their metabolic diversity and flexibility and are capable of metabolizing a variety of organic compounds, including complex carbohydrates, lipids, and proteins, under cold conditions. This metabolic flexibility allows them to sustain growth and survival in cold environments [54]. Studies have shown that as temperature decreases, *Acinetobacter* also becomes the dominant genus in the gut microbiota of *Carassius auratus* [20], further supporting its competitive advantage in cold environments. These microbial shifts reflect the selective pressure exerted by cold stress on the gut microbiota, leading to the redistribution of different bacterial species based on their metabolic traits and environmental adaptability [40]. In this study, the significant increase in *Acinetobacter* in the ULT group suggests its potential role in host nutrient metabolism under cold conditions, particularly in lipid and protein degradation [55]. This finding indicates that *Acinetobacter* may contribute to host energy acquisition in cold environments through metabolic adaptation.

In our dynamic temporal analysis, significant changes in the gut microbiota composition of both LT and ULT groups were observed as the duration of cold stress increased. α-Diversity analysis revealed that in the LT group, the Ace index at 0 h was significantly lower than at 12 h and 36 h, while the Shannon index remained unchanged across all time points. This suggests that although species richness initially declined under cold stress, overall microbial diversity remained stable. In contrast, in the more extreme cold conditions of the ULT group, the Ace index increased significantly at 36 h, whereas the Shannon index showed a marked decline at 24 h before recovering at 36 h. This pattern may indicate an initial phase of dysbiosis in response to cold stress, followed by the recolonization of bacteria better adapted to low temperatures [56]. Notably, in the ULT group, the stronger selective pressure exerted by cold stress likely reduced the abundance of temperature-sensitive bacterial taxa, leading to the observed decrease in the Shannon index at 24 h. However, by 36 h, microbial community restructuring occurred, with an expansion of cold-tolerant taxa, resulting in a significant increase in the Ace index and a subsequent recovery of Shannon diversity. This process may be driven by host-microbiota interactions, where cold stress influences the intestinal physiological environment, including mucus secretion, digestive enzyme activity, and immune modulation, thereby shaping the dynamic shifts in microbial composition [57]. In the LT group, there was a significant increase in the proportion of Firmicutes at 12 h, possibly related to short-term metabolic regulation in fish. Firmicutes are known for their robust metabolic adaptability, capable of efficiently utilizing complex carbon sources and other substrates through fermentation pathways to help maintain energy balance in cold environments [58]. This short-term enrichment suggests that they can rapidly respond to sudden temperature changes, thereby supporting the host's metabolic needs [59]. Additionally, the proportion of Verrucomicrobiota significantly increased after 36 h of cold stress, possibly reflecting their metabolic adaptation to nutritional limitations and cold conditions within the gut. Verrucomicrobiota are known for their unique capabilities in carbohydrate degradation and mucin utilization, effectively breaking down complex carbon sources at low temperatures and using host-secreted mucus as an energy source to maintain activity and proliferation under resource-limited conditions [60,61]. Cold environments often coincide with a reduction in the host's metabolic rate, leading to decreased nutrient availability in the gut, including carbohydrates and other energy sources [62]. Under such conditions, Verrucomicrobiota can exhibit a strong survival advantage by metabolizing complex sugars and utilizing host-secreted mucus [63]. In the

ULT group, the proportion of *Proteobacteria* decreased at 12 and 36 h, consistent with the trend observed in the LT group, suggesting that cold stress may have a similar inhibitory effect on Proteobacteria over time. Consistent with the LT group, the proportion of Firmicutes significantly increased at 12 h, indicating temporal differences in the adaptation of different microbial communities to cold stress. In colder conditions, the rapid enrichment of Firmicutes in the early stages of cold stress may be indicative of their adaptation to maintain the energy supply for the host in cold environments [64]. Meanwhile, a significant increase in Bacteroidota at 36 h suggests that they may require a longer period to adapt to cold conditions. These microbes are known for their ability to decompose complex polysaccharides and cellulose, often enriching the gut over extended periods of stress by efficiently utilizing limited resources [65].

At the genus level in the LT group, *Acinetobacter* and *Vibrio* were dominant at 0 h and 12 h, particularly *Vibrio*, which occupied a major proportion at 0 h but decreased over time. This suggests that the cold environment may inhibit the growth of *Vibrio* [66], while *Acinetobacter*, which has greater environmental adaptability, became the dominant genus at 36 h. Concurrently, *Mycoplasma* showed a significant increase at 12 h, suggesting a strong adaptive capacity to short-term cold exposure. Known for its streamlined metabolic pathways and tolerance to extreme environments, *Mycoplasma* can survive under low-nutrient conditions, a trait that may provide it with a competitive advantage during the early phase of cold stress [67]. As a result, *Mycoplasma* may act as a transiently responsive bacterial group, rapidly proliferating in the initial stages of cold exposure. However, as the cold stress persists, the host gradually adapts to environmental changes, leading to a stabilization of the gut microbiota structure. Consequently, *Mycoplasma* may be replaced by more stable bacterial communities. This dynamic shift suggests that certain bacterial groups may transiently expand in response to environmental stress. Yet, as ecological niches adjust and host adaptation strengthens, a more stable microbial community ultimately emerges [68,69]. In the ULT group, *Vibrio* was the dominant genus at 0 h, 12 h, and 24 h, with a sharp increase in abundance at 24 h, followed by a sharp decline at 36 h. This reflects *Vibrio*'s poor adaptation to even lower temperatures. Although *Vibrio* maintained a high abundance at 0 h, 12 h, and peaked at 24 h, indicating it could still sustain metabolic activity in short-term low-temperature environments [70], its ability to adapt diminishes as cold exposure persists. The drastic decline in Vibrio abundance at 36 h indicates that it may struggle to survive under prolonged low-temperature conditions, likely due to shifts in nutrient availability and host metabolic changes. In contrast, Acinetobacter exhibited a substantial increase at 36 h, emerging as the dominant genus, a pattern consistent with the LT group. Acinetobacter is well known for its metabolic versatility and strong environmental adaptability, allowing it to thrive in cold and nutrient-limited conditions [71,72]. Its replacement of Vibrio at 36 h may reflect the gut environment's further adaptation to prolonged cold stress, favoring the proliferation and colonization of more resilient bacterial communities. This shift also suggests that low temperatures not only alter the composition of the microbial community but may also influence its succession dynamics by modulating host metabolism and gut barrier function.

Sphingomonas and *Brevundimonas* both experienced a significant decline in abundance at 24 h. *Sphingomonas* is known for its capacity to degrade complex organic materials, particularly in environmental remediation [73]. However, its rate of complex organic matter decomposition may decrease under low temperatures, reducing its competitiveness [74]. Similarly, *Brevundimonas* has a wide distribution in various environments, but its metabolic efficiency may be limited under cold conditions, resulting in suppressed growth and reproduction [75]. Therefore, the decline of these two microbial groups may indicate weaker metabolic pathways and adaptability in cold environments, gradually being replaced by other microbial communities better adapted to prolonged low-temperature stress.

5. Conclusions

This study analyzed the gut microbiota changes in juvenile yellowfin tuna under cold stress, revealing the significant impacts of low temperatures on microbial diversity, composition, and metabolism. Results showed a gradual decrease in *Proteobacteria* abundance with lower temperatures, indicating weaker adaptability, while *Verrucomicrobiota*, *Firmicutes*, and *Bacteroidota* exhibited increases at 36 h, reflecting their stronger metabolic adaptability. Additionally, *Vibrio* abundance dropped significantly, suggesting poor adaptability to prolonged cold. Overall, this study provides new insights into the gut microbiota regulation of *Thunnus albacares* under low-temperature conditions, highlighting the metabolic adaptability of specific bacterial communities to cold stress. These findings have practical implications for aquaculture, aiding in optimizing feed formulations, probiotic supplementation, and temperature control to enhance fish health and growth. This study deepens our understanding of fish adaptation to extreme climates and offers valuable guidance for aquaculture management in cold environments.

Supplementary Materials: The following supporting information can be downloaded at: https://www.mdpi.com/article/10.3390/jmse13030602/s1, Table S1: 36 valid sequences of samples; Table S2: Table of sample coverage and diversity indices for the three treatment groups at different low-temperature stress time points.

Author Contributions: Conceptualization, J.H., Z.F. and T.Z.; Methodology, J.H. and Z.F.; Software, J.W. and B.H.; Validation, J.H., Z.F., W.Y. and B.H.; Formal analysis, J.H. and W.Y.; Investigation, J.H. and T.Z.; Resources, J.W. and Z.M.; Data curation, J.H. and B.H.; Writing—original draft, J.H.; Writing—review & editing, Z.F. and Z.M.; Visualization, J.H. and T.Z.; Supervision, Z.M.; Project administration, J.W. and W.Y.; Funding acquisition, J.W., W.Y. and Z.M. All authors have read and agreed to the published version of the manuscript.

Funding: This work was supported by the Research on Breeding Technology of Candidate Species for Guangdong Modern Marine Ranching (2024-MRB-00-001); the National Natural Science Foundation of China (32460927); the Science and Technology special fund of Hainan Province (ZDYF2024XDNY247); and the Central Public-interest Scientific Institution Basal Research Fund, CAFS (2024XT04, 2023TD58). The research was supported by the Project of Sanya Yazhou Bay Science and Technology City (SKJC-2022-PTDX-015).

Institutional Review Board Statement: The experiments were in compliance with the regulations and guidelines established by the Animal Care and Use Committee of South China Sea Fisheries Research Institute, Chinese Academy of Fishery Sciences, and permitted by this committee.

Data Availability Statement: The data presented in this study are openly available in NCBI Sequence Read Archive at https://www.ncbi.nlm.nih.gov/bioproject/PRJNA1190122/ (accessed on 24 November 2024), reference number PRJNA1190122.

Conflicts of Interest: Author Bowen Hou was employed by Hainan Haiken Marine Industry Group Co., Ltd. Author Jinhui Wu was employed by Agro-Tech Extension Center of Guangdong Province. Author Tao Zhang was employed by Dalian Tianzheng Industrial Co., Ltd. The remaining authors declare that the research was conducted in the absence of any commercial or financial relationships that could be construed as a potential conflict of interest.

References

1. Merrifield, D.L.; Rodiles, A. The Fish Microbiome and Its Interactions with Mucosal Tissues. In *Mucosal Health in Aquaculture*; Elsevier: Amsterdam, The Netherlands, 2015; pp. 273–295.
2. Banerjee, G.; Ray, A.K. Bacterial Symbiosis in the Fish Gut and Its Role in Health and Metabolism. *Symbiosis* **2017**, *72*, 1–11. [CrossRef]
3. Sehnal, L.; Brammer-Robbins, E.; Wormington, A.M.; Blaha, L.; Bisesi, J.; Larkin, I.; Martyniuk, C.J.; Simonin, M.; Adamovsky, O. Microbiome Composition and Function in Aquatic Vertebrates: Small Organisms Making Big Impacts on Aquatic Animal Health. *Front. Microbiol.* **2021**, *12*, 567408. [CrossRef]
4. Romero, J.; Ringø, E.; Merrifield, D.L. The Gut Microbiota of Fish. In *Aquaculture Nutrition: Gut Health, Probiotics and Prebiotics*; John Wiley & Sons, Ltd.: Hoboken, NJ, USA, 2014; pp. 75–100.
5. Altringham, J.D.; Block, B.A. Why Do Tuna Maintain Elevated Slow Muscle Temperatures? Power Output of Muscle Isolated from Endothermic and Ectothermic Fish. *J. Exp. Biol.* **1997**, *200*, 2617–2627. [CrossRef]
6. Stevens, E.D.; Neill, W.H. Body Temperature Relations of Tunas, Especially Skipjack. In *Fish Physiology*; Elsevier: Amsterdam, The Netherlands, 1978; pp. 315–359.
7. Zhang, L.; Yang, Z.; Yang, F.; Wang, G.; Zeng, M.; Zhang, Z.; Yang, M.; Wang, Z.; Li, Z. Gut Microbiota of Two Invasive Fishes Respond Differently to Temperature. *Front. Microbiol.* **2023**, *14*, 1087777. [CrossRef] [PubMed]
8. Sepulveda, J.; Moeller, A.H. The Effects of Temperature on Animal Gut Microbiomes. *Front. Microbiol.* **2020**, *11*, 384. [CrossRef]
9. Wang, B.; Zhang, S.-Q.; Dong, J.-L.; Li, Y.; Jin, Y.-X.; Xiao, H.-W.; Wang, H.-C.; Fan, S.-J.; Cui, M. Ambient Temperature Structures the Gut Microbiota of Zebrafish to Impact the Response to Radioactive Pollution. *Environ. Pollut.* **2022**, *293*, 118539. [CrossRef] [PubMed]
10. Li, P.; Zhang, J.; Liu, X.; Gan, L.; Xie, Y.; Zhang, H.; Si, J. The Function and the Affecting Factors of the Zebrafish Gut Microbiota. *Front. Microbiol.* **2022**, *13*, 903471. [CrossRef]
11. Block, B.A.; Keen, J.E.; Castillo, B.; Dewar, H.; Freund, E.V.; Marcinek, D.J.; Brill, R.W.; Farwell, C. Environmental Preferences of Yellowfin Tuna (*Thunnus albacares*) at the Northern Extent of Its Range. *Mar. Biol.* **1997**, *130*, 119–132. [CrossRef]
12. Artetxe-Arrate, I.; Fraile, I.; Marsac, F.; Farley, J.H.; Rodriguez-Ezpeleta, N.; Davies, C.R.; Clear, N.P.; Grewe, P.; Murua, H. A Review of the Fisheries, Life History and Stock Structure of Tropical Tuna (Skipjack *Katsuwonus pelamis*, Yellowfin *Thunnus albacares* and Bigeye *Thunnus obesus*) in the Indian Ocean. *Adv. Mar. Biol.* **2021**, *88*, 39–89.
13. Lan, K.-W.; Evans, K.; Lee, M.-A. Effects of Climate Variability on the Distribution and Fishing Conditions of Yellowfin Tuna (*Thunnus albacares*) in the Western Indian Ocean. *Clim. Change* **2013**, *119*, 63–77. [CrossRef]
14. Lilly, L.E.; Bonaventura, J.; Lipnick, M.S.; Block, B.A. Effect of Temperature Acclimation on Red Blood Cell Oxygen Affinity in Pacific Bluefin Tuna (*Thunnus orientalis*) and Yellowfin Tuna (*Thunnus albacares*). *Comp. Biochem. Physiol. Part A Mol. Integr. Physiol.* **2015**, *181*, 36–44. [CrossRef] [PubMed]
15. Sun, W.; Zhang, J.; Meng, J.; Liu, Y. Sea Surface Temperature Characteristics and Trends in China Offshore Seas from 1982 to 2017. *J. Coast. Res.* **2019**, *90*, 27–34. [CrossRef]
16. Schwing, F.B.; Murphree, T.; deWitt, L.; Green, P.M. The Evolution of Oceanic and Atmospheric Anomalies in the Northeast Pacific During the El Niño and La Niña Events of 1995–2001. *Prog. Oceanogr.* **2002**, *54*, 459–491. [CrossRef]
17. Pörtner, H.-O.; Berdal, B.; Blust, R.; Brix, O.; Colosimo, A.; De Wachter, B.; Giuliani, A.; Johansen, T.; Fischer, T.; Knust, R.; et al. Climate Induced Temperature Effects on Growth Performance, Fecundity and Recruitment in Marine Fish: Developing a Hypothesis for Cause and Effect Relationships in Atlantic Cod (*Gadus morhua*) and Common Eelpout (*Zoarces viviparus*). *Cont. Shelf Res.* **2001**, *21*, 1975–1997. [CrossRef]
18. McKenzie, D.J.; Axelsson, M.; Chabot, D.; Claireaux, G.; Cooke, S.J.; Corner, R.A.; De Boeck, G.; Domenici, P.; Guerreiro, P.M.; Hamer, B.; et al. Conservation Physiology of Marine Fishes: State of the Art and Prospects for Policy. *Conserv. Physiol.* **2016**, *4*, cow046. [CrossRef]
19. Crawshaw, L.I. Low-Temperature Dormancy in Fish. *Am. J. Physiol.-Regul. Integr. Comp. Physiol.* **1984**, *246*, R479–R486. [CrossRef]
20. Liu, Y.; Lv, H.; Xu, L.; Zhang, K.; Mei, Y.; Chen, J.; Wang, M.; Guan, Y.; Pang, H.; Wang, Y.; et al. The Effect of Dietary Lactic Acid Bacteria on Intestinal Microbiota and Immune Responses of Crucian Carp (*Carassius auratus*) under Water Temperature Decrease. *Front. Microbiol.* **2022**, *13*, 847167. [CrossRef] [PubMed]
21. Lv, H.; Liu, Y.; Li, H.; Yin, X.; Wang, P.; Qu, X.; Gao, Y.; Li, W.; Chu, Z. Modulation of Antioxidant Enzymes, Heat Shock Protein, and Intestinal Microbiota of Large Yellow Croaker (*Larimichthys crocea*) under Acute Cold Stress. *Front. Mar. Sci.* **2021**, *8*, 725899. [CrossRef]
22. Liu, C.; Zhao, L.-P.; Shen, Y.-Q. A Systematic Review of Advances in Intestinal Microflora of Fish. *Fish Physiol. Biochem.* **2021**, *47*, 2041–2053. [CrossRef]
23. Chen, C.-Z.; Li, P.; Liu, L.; Li, Z.-H. Exploring the Interactions between the Gut Microbiome and the Shifting Surrounding Aquatic Environment in Fisheries and Aquaculture: A Review. *Environ. Res.* **2022**, *214*, 114202. [CrossRef]

24. Bakke, A.M.; Glover, C.; Krogdahl, Å. Feeding, Digestion and Absorption of Nutrients. In *Fish Physiology*; Elsevier: Amsterdam, The Netherlands, 2010; pp. 57–110.
25. Ray, A.K.; Ringø, E. The Gastrointestinal Tract of Fish. In *Aquaculture Nutrition: Gut Health, Probiotics and Prebiotics*; John Wiley & Sons, Ltd.: Hoboken, NJ, USA, 2014; pp. 1–13.
26. Lobionda, S.; Sittipo, P.; Kwon, H.Y.; Lee, Y.K. The Role of Gut Microbiota in Intestinal Inflammation with Respect to Diet and Extrinsic Stressors. *Microorganisms* **2019**, *7*, 271. [CrossRef]
27. Gasaly, N.; de Vos, P.; Hermoso, M.A. Impact of Bacterial Metabolites on Gut Barrier Function and Host Immunity: A Focus on Bacterial Metabolism and Its Relevance for Intestinal Inflammation. *Front. Immunol.* **2021**, *12*, 658354. [CrossRef]
28. Hansen, G.H.; Olafsen, J.A. Bacterial Interactions in Early Life Stages of Marine Cold Water Fish. *Microb. Ecol.* **1999**, *38*, 1–26. [CrossRef] [PubMed]
29. Yan, H.; Soon, W.; Wang, Y. A Composite Sea Surface Temperature Record of the Northern South China Sea for the Past 2500 Years: A Unique Look into Seasonality and Seasonal Climate Changes During Warm and Cold Periods. *Earth-Sci. Rev.* **2015**, *141*, 122–135. [CrossRef]
30. Cakin, I.; Morrissey, B.; Gordon, M.; Gaffney, P.P.; Marcello, L.; Macgregor, K.; Taggart, M.A. Comparing DNA Isolation and Sequencing Strategies for 16S rRNA Gene Amplicon Analysis in Biofilm Containing Environments. *J. Microbiol. Methods* **2024**, *220*, 106921. [CrossRef] [PubMed]
31. Schloss, P.D.; Westcott, S.L.; Ryabin, T.; Hall, J.R.; Hartmann, M.; Hollister, E.B.; Lesniewski, R.A.; Oakley, B.B.; Parks, D.H.; Robinson, C.J.; et al. Introducing Mothur: Open-Source, Platform-Independent, Community-Supported Software for Describing and Comparing Microbial Communities. *Appl. Environ. Microbiol.* **2009**, *75*, 7537–7541. [CrossRef] [PubMed]
32. Edgar, R.C.; Haas, B.J.; Clemente, J.C.; Quince, C.; Knight, R. Uchime Improves Sensitivity and Speed of Chimera Detection. *Bioinformatics* **2011**, *27*, 2194–2200. [CrossRef]
33. Lan, Y.; Wang, Q.; Cole, J.R.; Rosen, G.L. Using the Rdp Classifier to Predict Taxonomic Novelty and Reduce the Search Space for Finding Novel Organisms. *PLoS ONE* **2012**, *7*, e32491. [CrossRef]
34. Koetschan, C.; Kittelmann, S.; Lu, J.; Al-Halbouni, D.; Jarvis, G.N.; Müller, T.; Wolf, M.; Janssen, P.H. Internal Transcribed Spacer 1 Secondary Structure Analysis Reveals a Common Core Throughout the Anaerobic Fungi (Neocallimastigomycota). *PLoS ONE* **2014**, *9*, e91928. [CrossRef]
35. Kang, X.; Liu, G.; Liu, Y.; Xu, Q.; Zhang, M.; Fang, M. Transcriptome Profile at Different Physiological Stages Reveals Potential Mode for Curly Fleece in Chinese Tan Sheep. *PLoS ONE* **2013**, *8*, e71763. [CrossRef]
36. Hong, J.; Chen, M.; Deng, Z.; Li, Y.; Wang, Y. Variance Analysis of Intestinal Microbial Diversity of the Noble Scallop (*Chlamys nobilis*) under Enrofloxacin Exposure. *Pak. J. Zool.* **2021**, *53*, 1221–1231. [CrossRef]
37. Ming, J.; Fu, Z.; Ma, Z.; Zhou, L.; Zhang, Z.; Song, C.; Yuan, X.; Wu, Q. The Effect of Sulfamonomethoxine Treatment on the Gut Microbiota of Nile Tilapia (*Oreochromis niloticus*). *MicrobiologyOpen* **2020**, *9*, e1116. [CrossRef]
38. Kim, J.A.; Park, Y.-S.; Kim, J.-H.; Choi, C.Y. Impact of Water Temperature on Oxidative Stress and Intestinal Microbiota in Pearl-Spot Chromis, *Chromis notata* (Temminck & Schlegel, 1843). *Comp. Biochem. Physiol. Part B Biochem. Mol. Biol.* **2024**, *275*, 111029.
39. Li, R.-X.; Amenyogbe, E.; Lu, Y.; Jin, J.-H.; Xie, R.-T.; Huang, J.-S. Effects of Low-Temperature Stress on Serum Biochemical Indicators, Intestinal Microbiome, and Transcriptome of Juvenile Golden Pompano (*Trachinotus ovatus*). *Aquac. Int.* **2024**, *32*, 5551–5578. [CrossRef]
40. Hu, J.; Zhao, H.; Wang, G.; Sun, Y.; Wang, L. Energy Consumption and Intestinal Microbiome Disorders of Yellow Catfish (*Pelteobagrus fulvidraco*) under Cold Stress. *Front. Physiol.* **2022**, *13*, 985046. [CrossRef]
41. Ma, S.; Lv, Y.; Hou, L.; Jia, Z.; Lin, S.; Wang, S.; He, X.; Hou, J. Effect of Acute Temperature Stress on Energy Metabolism, Immune Performance and Gut Microbiome of Largemouth Bass (*Micropterus salmoides*). *Aquac. Fish.* 2023, in press. [CrossRef]
42. Murashita, K.; Hashimoto, H.; Takashi, T.; Eba, T.; Kumon, K.; Matsunari, H.; Soma, S.; Oku, H.; Furuita, H.; Yoshinaga, H.; et al. Characterization of Digestive Physiology in Pacific Bluefin Tuna *Thunnus orientalis* juveniles fed a raw fish feed and a commercial diet. *Aquaculture* **2021**, *538*, 736562. [CrossRef]
43. Villarini, G.; Vecchi, G.A. Multiseason Lead Forecast of the North Atlantic Power Dissipation Index (PDI) and Accumulated Cyclone Energy (ACE). *J. Clim.* **2013**, *26*, 3631–3643. [CrossRef]
44. Li, R.-X.; Amenyogbe, E.; Lu, Y.; Jin, J.-H.; Xie, R.-T.; Huang, J.-S. Effects of Low-Temperature Stress on Intestinal Structure, Enzyme Activities and Metabolomic Analysis of Juvenile Golden Pompano (*Trachinotus ovatus*). *Front. Mar. Sci.* **2023**, *10*, 1114120. [CrossRef]
45. Liu, Y.; Cheng, J.; Xia, Y.; Li, X.; Liu, Y.; Liu, P.-F. Response Mechanism of Gut Microbiome and Metabolism of European Seabass (*Dicentrarchus labrax*) to Temperature Stress. *Sci. Total Environ.* **2022**, *813*, 151786. [CrossRef]
46. Huus, K.E.; Ley, R.E. Blowing Hot and Cold: Body Temperature and the Microbiome. *mSystems* **2021**, *6*, e0070721. [CrossRef] [PubMed]

47. Zhou, C.; Yang, S.; Ka, W.; Gao, P.; Li, Y.; Long, R.; Wang, J. Association of Gut Microbiota with Metabolism in Rainbow Trout under Acute Heat Stress. *Front. Microbiol.* **2022**, *13*, 846336. [CrossRef] [PubMed]
48. Yoon, D.-S.; Kim, D.-H.; Kim, J.-H.; Sakakura, Y.; Hagiwara, A.; Park, H.G.; Lee, M.-C.; Lee, J.-S. Interactions Between Lipid Metabolism and the Microbiome in Aquatic Organisms: A Review. *Mar. Pollut. Bull.* **2024**, *207*, 116858. [CrossRef] [PubMed]
49. Huang, K.; He, Y.; Wang, W.; Jiang, R.; Zhang, Y.; Li, J.; Zhang, X.-X.; Wang, D. Temporal Differentiation in the Adaptation of Functional Bacteria to Low-Temperature Stress in Partial Denitrification and Anammox System. *Environ. Res.* **2024**, *244*, 117933. [CrossRef]
50. Quatrini, R.; Johnson, D.B. Microbiomes in Extremely Acidic Environments: Functionalities and Interactions That Allow Survival and Growth of Prokaryotes at Low pH. *Curr. Opin. Microbiol.* **2018**, *43*, 139–147. [CrossRef]
51. Reineke, W.; Schlömann, M. Carbon Cycle. In *Environmental Microbiology*; Springer: Berlin/Heidelberg, Germany, 2023; pp. 71–126.
52. Wasmund, K.; Pelikan, C.; Schintlmeister, A.; Wagner, M.; Watzka, M.; Richter, A.; Bhatnagar, S.; Noel, A.; Hubert, C.R.J.; Rattei, T. Genomic Insights into Diverse Bacterial Taxa That Degrade Extracellular DNA in Marine Sediments. *Nat. Microbiol.* **2021**, *6*, 885–898. [CrossRef]
53. Sun, Y.; Zhang, S.; Nie, Q.; He, H.; Tan, H.; Geng, F.; Ji, H.; Hu, J.; Nie, S. Gut Firmicutes: Relationship with Dietary Fiber and Role in Host Homeostasis. *Crit. Rev. Food Sci. Nutr.* **2023**, *63*, 12073–12088. [CrossRef]
54. Towner, K.J. The Genus Acinetobacter. In *The Prokaryotes: A Handbook on the Biology of Bacteria: Ecophysiology, Isolation, Identification, Applications*; Springer: Berlin/Heidelberg, Germany, 1992; pp. 3137–3143.
55. Wang, X.-Y.; Xie, J. Characterization of Metabolite, Genome and Volatile Organic Compound Changes Provides Insights into the Spoilage and Cold Adaptive Markers of Acinetobacter Johnsonii XY27. *LWT* **2022**, *162*, 113453. [CrossRef]
56. Yi, W.; Cheng, J.; Wei, Q.; Pan, R.; Song, S.; He, Y.; Tang, C.; Liu, X.; Zhou, Y.; Su, H. Effect of Temperature Stress on Gut-Brain Axis in Mice: Regulation of Intestinal Microbiome and Central Nlrp3 Inflammasomes. *Sci. Total Environ.* **2021**, *772*, 144568. [CrossRef]
57. Sanahuja, I.; Fernández-Alacid, L.; Sánchez-Nuño, S.; Ordóñez-Grande, B.; Ibarz, A. Chronic Cold Stress Alters the Skin Mucus Interactome in a Temperate Fish Model. *Front. Physiol.* **2019**, *9*, 1916. [CrossRef]
58. Zhang, X.-Y.; Sukhchuluun, G.; Bo, T.-B.; Chi, Q.-S.; Yang, J.-J.; Chen, B.; Zhang, L.; Wang, D.-H. Huddling Remodels Gut Microbiota to Reduce Energy Requirements in a Small Mammal Species During Cold Exposure. *Microbiome* **2018**, *6*, 103.
59. Zhang, X.-Y.; Wang, D.-H. Gut Microbial Community and Host Thermoregulation in Small Mammals. *Front. Physiol.* **2022**, *13*, 888324. [CrossRef]
60. Pan, Y.; Sun, R.-Z.; Wang, Y.; Chen, G.-L.; Fu, Y.-Y.; Yu, H.-Q. Carbon Source Shaped Microbial Ecology, Metabolism and Performance in Denitrification Systems. *Water Res.* **2023**, *243*, 120330. [CrossRef] [PubMed]
61. Panah, F.M.; Lauridsen, C.; Højberg, O.; Jensen, H.E.; Nielsen, T.S. Composition of Mucus-and Digesta-Associated Bacteria in Growing Pigs with and without Diarrhea Differed According to the Presence of Colonic Inflammation. *BMC Microbiol.* **2023**, *23*, 145. [CrossRef]
62. Carey, H.V.; Assadi-Porter, F.M. The Hibernator Microbiome: Host-Bacterial Interactions in an Extreme Nutritional Symbiosis. *Annu. Rev. Nutr.* **2017**, *37*, 477–500. [CrossRef] [PubMed]
63. Zhang, F.; Zhao, Y.; Liu, X.; Li, Z.; Liu, N.; Tang, L.; Jiang, Q.; Fan, Z.; Tan, B.; Li, Y.; et al. Effects of Soluble Glucomannan and Insoluble Cellulose Treatment on Mucin Secretion and Mucin Glycosylation-Related Gene Expression in the Colons of Mice. *J. Sci. Food Agric.* **2023**, *103*, 7739–7746. [CrossRef]
64. Karl, J.P.; Hatch, A.M.; Arcidiacono, S.M.; Pearce, S.C.; Pantoja-Feliciano, I.G.; Doherty, L.A.; Soares, J.W. Effects of Psychological, Environmental and Physical Stressors on the Gut Microbiota. *Front. Microbiol.* **2018**, *9*, 2013. [CrossRef]
65. Ma, X.; La, Y.; Yang, G.; Dai, R.; Zhang, J.; Zhang, Y.; Jin, J.; Ma, X.; Guo, X.; Chu, M.; et al. Multi-Omics Revealed the Effects of Dietary Energy Levels on the Rumen Microbiota and Metabolites in Yaks under House-Feeding Conditions. *Front. Microbiol.* **2024**, *14*, 1309535. [CrossRef]
66. Biosca, E.G.; Amaro, C.; Marco-Noales, E.; Oliver, J.D. Effect of Low Temperature on Starvation-Survival of the Eel Pathogen Vibrio Vulnificus Biotype 2. *Appl. Environ. Microbiol.* **1996**, *62*, 450–455. [CrossRef]
67. Tardy, F.; Mick, V.; Dordet-Frisoni, E.; Marenda, M.S.; Sirand-Pugnet, P.; Blanchard, A.; Citti, C. Integrative Conjugative Elements Are Widespread in Field Isolates of Mycoplasma Species Pathogenic for Ruminants. *Appl. Environ. Microbiol.* **2015**, *81*, 1634–1643. [CrossRef]
68. Kaboré, O.D.; Godreuil, S.; Drancourt, M. Planctomycetes as Host-Associated Bacteria: A Perspective That Holds Promise for Their Future Isolations, by Mimicking Their Native Environmental Niches in Clinical Microbiology Laboratories. *Front. Cell. Infect. Microbiol.* **2020**, *10*, 519301. [CrossRef] [PubMed]
69. Gaca, A.O.; Lemos, J.A. Adaptation to Adversity: The Intermingling of Stress Tolerance and Pathogenesis in Enterococci. *Microbiol. Mol. Biol. Rev.* **2019**, *83*, e00074-18. [CrossRef] [PubMed]
70. Singleton, F.L.; Attwell, R.; Jangi, S.; Colwell, R.R. Effects of Temperature and Salinity on Vibrio Cholerae Growth. *Appl. Environ. Microbiol.* **1982**, *44*, 1047–1058. [CrossRef] [PubMed]

71. Percival, S.L.; Yates, M.V.; Williams, D.W.; Chalmers, R.M.; Gray, N.F. Acinetobacter. In *Microbiology of Waterborne Diseases*; Elsevier: Amsterdam, The Netherlands, 2014; pp. 35–48.
72. Kour, D.; Yadav, A.N. Alleviation of Cold Stress in Wheat with Psychrotrophic Phosphorus Solubilizing Acinetobacter Rhizosphaerae Eu-Kl44. *Braz. J. Microbiol.* **2023**, *54*, 371–383. [CrossRef]
73. Asaf, S.; Numan, M.; Khan, A.L.; Al-Harrasi, A. *Sphingomonas*: From Diversity and Genomics to Functional Role in Environmental Remediation and Plant Growth. *Crit. Rev. Biotechnol.* **2020**, *40*, 138–152. [CrossRef]
74. Yi, X.; Guo, H.; Deng, J.; Bai, Z.J.; Shi, X.F.; Qiang, J.B. Effect of *Sphingomonas polyaromaticivorans* on Spontaneous Combustion of Lignite During Low-Temperature Oxidation. *Fuel* **2024**, *377*, 132786. [CrossRef]
75. Deshpande, N.M.; Sarnaik, S.S.; Paranjpe, S.A.; Kanekar, P.P. Optimization of Dimethoate Degradation by *Brevundimonas* sp. MCM B-427 Using Factorial Design: Studies on Interactive Effects of Environmental Factors. *World J. Microbiol. Biotechnol.* **2004**, *20*, 455–462. [CrossRef]

Journal of
Marine Science and Engineering

MDPI

Article

Effects of Different Cryoprotectants on Cryopreservation of Sperm from Noble Scallop *Mimachlamys nobilis*

Minghao Li [1,2,3,4,5], Zhengyi Fu [1,2,3,4,6], Gang Yu [1,2,3,4,*] and Zhenhua Ma [1,2,3,4,6,*]

1 Key Laboratory of Efficient Utilization and Processing of Marine Fishery Resources of Hainan Province, Sanya Tropical Fisheries Research Institute, Sanya 572018, China
2 South China Sea Fisheries Research Institute, Chinese Academy of Fishery Sciences, Guangzhou 510300, China
3 Hainan Engineering Research Center for Deep-Sea Aquaculture and Processing, Sanya Tropical Fisheries Research Institute, Sanya 572018, China
4 International Joint Research Center for Conservation and Application of Fishery Resources in the South China Sea, Sanya Tropical Fisheries Research Institute, Sanya 572018, China
5 College of Fisheries, Tianjin Agricultural University, Tianjin 300384, China
6 College of Science and Engineering, Flinders University, Adelaide 5001, Australia
* Correspondence: yugang@scsfri.ac.cn (G.Y.); zhenhua.ma@scsfri.ac.cn (Z.M.); Tel.: +86-13360556962 (G.Y.); +86-13168882678 (Z.M.)

Abstract: In cryopreservation technology, the choice of cryoprotectant plays a crucial role in cell survival and function. Different types of cryoprotectants, each with unique protective mechanisms, mitigate cellular damage from ice crystal formation during freezing. This study investigated the effects of different types and concentrations of cryoprotectants on the cryopreservation efficacy of noble scallop *Mimachlamys nobilis* sperm. Six cryoprotectants were tested, including four permeable cryoprotectants (dimethyl sulfoxide (DMSO), ethylene glycerol (EG), propylene glycerol (PG), methanol (MET)) and two non-permeable cryoprotectants (trehalose (TRE), fetal bovine serum (FBS)). The results showed that permeable cryoprotectants, which penetrate the cell membrane, regulate the osmotic pressure inside and outside cells to reduce dehydration damage. Among them, 10% DMSO provided the best protection, significantly preserving sperm motility, velocity, and morphology. Non-permeable cryoprotectants, although unable to penetrate cells, stabilized the extracellular environment at higher concentrations (such as FBS). Additionally, MET and FBS exhibited enhanced protective effects with increasing concentration, indicating their potential in reducing sperm structural damage at higher concentrations. Morphological observations indicated that freezing caused varying degrees of structural damage to sperm, with flagellar integrity being crucial for motility. Overall, selecting an appropriate cryoprotectant and concentration is essential for the efficient cryopreservation of *M. nobilis* sperm, providing a valuable reference for conserving germplasm resources of marine species.

Keywords: *Mimachlamys nobilis*; cryoprotective agent; sperm motility; sperm velocity; sperm morphology; ultrastructure

1. Introduction

The noble scallop (*Mimachlamys nobilis*), classified under the phylum Mollusca, class Lamellibranchia, order Pterioida, and family Pectinidae1, is recognized for its relatively large size and swift growth cycle [1]. Its adductor muscle, often referred to as "dried scallop", holds substantial commercial importance [2]. In China, the cultivation of scallops has become a pillar industry driving coastal economic growth, with scallops standing among the country's most prominent seafood exports [3]. Geographically, this species is chiefly found in Japan, in Indonesia, and along China's southeastern shoreline [4]. Remarkably, *M. nobilis* sperm remain motile for over 40 h after activation by natural seawater, exhibiting high swimming speeds, earning them the designation "marathon-type sperm" [5].

In order to promote effective reproduction and enhance the economic benefits of *M. nobilis*, research on and the optimization of ultralow-temperature cryopreservation techniques for its sperm are critically important. Cryobiology was initiated in 1949 when Polge and Smith discovered that glycerol could be used for the low-temperature preservation of animal sperm [6]. Cryobiological principles suggest that nearly all biochemical activities in living organisms are enzyme-dependent, with enzyme activity decreasing progressively with temperature reduction [7]. Below a critical temperature, enzyme activity ceases temporarily, halting all metabolic processes and inducing a low-energy dormant state that enables long-term preservation [8]. Upon thawing, these processes can be reactivated. Cryopreservation at ultralow temperatures not only extends the storage time of sperm but also improves the success rate of artificial insemination [9]. However, while most studies have focused on optimal freezing temperatures and cooling rates, the selection and formulation of effective cryoprotectants and their concentrations remain unresolved challenges. Further research is needed to optimize cryoprotectant types and dosages to minimize cellular damage during freezing and thawing, thereby preserving sperm viability and motility.

Cryoprotectants play a crucial role in minimizing cryoinjury during cryopreservation by protecting the structural and functional integrity of cells. Based on their permeability to cell membranes, cryoprotectants are classified as permeating or non-permeating [10]. Permeating cryoprotectants exhibit high membrane penetration capabilities, allowing them to interact with intracellular water, reduce the concentration of electrolytes in unfrozen solutions, lower the freezing point, and inhibit ice crystal formation [11]. Their protective effect generally increases with concentration; however, at higher concentrations, they can exhibit cytotoxicity, necessitating the careful determination of safe concentration ranges [12]. Ideal permeating cryoprotectants possess high permeability and low toxicity. Common permeating cryoprotectants used in the cryopreservation of marine germplasm include dimethyl sulfoxide (DMSO), ethylene glycol (EG), propylene glycol (PG), and methanol (MET) [13]. Among these, DMSO is recognized as an effective cryoprotectant and is commonly used for bivalve sperm cryopreservation at concentrations between 5 and 15% [14].

Non-permeating cryoprotectants, on the other hand, excel at stabilizing cell membranes during freezing. They regulate the movement of intracellular and extracellular contents, maintaining the integrity and functionality of the membrane [15]. Common non-permeating cryoprotectants include trehalose (TRE), fetal bovine serum (FBS), glucose (GLC), polyethylene glycol (PEG), and fructose (FRU) [16]. Sugars interact with lipid structures on cell membranes, stabilizing and protecting their structure, while FBS mitigates the adverse effects of permeating ions. Non-permeating cryoprotectants can be used individually or in combination with permeating cryoprotectants [17]. However, there is limited research on cryoprotectants for *M. nobilis* sperm.

The effectiveness of cryoprotectant types and concentrations varies across species, which is a major reason for the lack of standardized cryoprotectant protocols. DMSO, MET, and EG are commonly used as effective permeating cryoprotectants in the cryopreservation of Pacific oyster (*Crassostrea gigas*) sperm [18]. DMSO and PG have been found suitable for the cryopreservation of Kumamoto oyster (*Crassostrea sikamea*) sperm [19]. A solution comprising 10–15% DMSO, 5% sucrose, and 10% egg yolk has been reported to be effective for the cryopreservation of Yesso scallop (*Patinopecten yessoensis*) sperm [20]. Additionally, mixtures of DMSO and various non-permeating cryoprotectants (e.g., FBS) have been successfully used in the cryopreservation of black-lip pearl oyster (*Pinctada margaritifera*) sperm [21].

This study aims to investigate the effects of different cryoprotectants and their concentrations on sperm motility, velocity, and ultrastructure in *M. nobilis*. By comprehensively assessing these parameters, this research seeks to reveal the physiological state of, damage to, and impact on sperm following ultralow-temperature freezing and thawing. These findings are anticipated to expand our understanding of cryobiology and reproductive biology in marine organisms, while offering scientific and technical support to address reproductive cycle constraints in artificial insemination. Ultimately, this could lead to higher scallop yields and greater economic returns.

2. Materials and Methods

2.1. Materials

In January 2024, *M. nobilis* specimens were obtained from Xincun Harbor in Lingshui Li Autonomous County, Hainan Province, China. Throughout the study, strict attention was given to stabilizing key environmental parameters—particularly temperature and salinity—to ensure the reliability of the experiment. During specimen collection, the water temperature remained at 25 ± 3 °C, while salinity was held at 35 ± 0.2 ‰, with both measurements continuously tracked by a PR730 series precision digital thermometer (Taian Panran Measurement and Control Technology Co., Ltd., Taian, China) and a SALTTEST11 meter from Eutech™ (Thermo Fisher Scientific, Waltham, MA, USA). An HK-HWJ-7200 aquaculture incubator (Guangzhou Huankong Agricultural Biotechnology Co., Ltd., Guangzhou, China) was used to maintain consistent temperatures throughout the experimental period, thus reducing environmental fluctuation impacts on sperm quality. Additionally, salinity was kept steady during sample processing to preserve uniform conditions.

2.2. Trait Data Collection

From a total of seventy *M. nobilis* specimens, three males displaying the most pronounced gonadal development were chosen. Their gonads were visibly enlarged, were milky white, and showed branching ducts, indicative of maturity. An examination of sperm development confirmed that these individuals were at an optimal reproductive stage (Table 1).

Table 1. Experimental animal data.

Species	Shell Length (mm)	Wet Weight (g)	Gonad Weight (g)
M. nobiliss 1	55.63	56.24	15.74
M. nobiliss 2	57.85	60.42	16.52
M. nobiliss 3	62.46	62.76	16.43

Note: actual measurement data to be filled in based on experimental results.

2.3. Sperm Collection

Among the seventy *M. nobilis* specimens, three males with the most developed gonads were chosen. To expose these gonads, a sterile scalpel was used to incise one side of the adductor muscle, revealing a visceral mass largely covered by white or pale-yellow gonadal tissue. The gonads were then carefully separated, gently blotted with clean absorbent paper to remove residual seawater or tissue fluids, and subsequently weighed. Prior to sperm collection, all instruments (including collection tubes and pipettes) underwent either high-temperature sterilization or disinfection with 70% ethanol to prevent external contamination. Additionally, work surfaces and equipment were disinfected in a clean laboratory environment to minimize microbial interference.

During sample collection, operators wore disposable gloves and masks to prevent any direct contact between their skin or respiratory system and the specimens. The excised gonads were placed on a 450-mesh sieve cloth and finely minced with scissors before being rinsed in precooled Hank's solution (supplied by Shantou Xilong Scientific Co., Ltd., Shantou, China; 99% analytical grade). Finally, the samples were maintained at 4 °C to preserve sperm quality and minimize external environmental influences.

Subsequently, the sperm obtained from these three individuals were pooled to form a homogeneous composite sample as the singular experimental unit for all subsequent assays.

2.4. *Sperm Density Determination and Evaluation of Fresh Sperm Viability and Motility*

During the activation phase, filtered natural seawater (0.25 μm pore size) served as the medium. The sperm suspension was combined with this seawater at a 1:10 ratio in a 15 mL conical tube and then vigorously agitated using a basic vortex oscillator (Thermo Fisher Scientific). A 5 μL aliquot of the resulting mixture was transferred to a sperm counting slide, where sperm activity was monitored via a Beionmed S3 analyzer (Shanghai Beion Medical Ltd., Shanghai, China) until most sperm ceased movement. All procedures were conducted at room temperature (25 ± 2 °C) on a 23 °C temperature-controlled platform. The analyzer provided measurements of sperm density, fresh sperm viability, and motility.

Subsequently, the sperm obtained from these three individuals were pooled to form a homogeneous composite sample as the singular experimental unit for all subsequent assays.

M. nobilis specimens with sperm motility exceeding 50% at 30 s post-activation were selected for further experiments. Each sample underwent three parallel tests, with three reliable data points recorded in each trial.

2.5. *Cryoprotectant Solution Preparation*

The diluent used in this experiment was selected through multiple preliminary trials, which identified Hank's solution as the optimal diluent. As cryoprotectants, dimethyl sulfoxide (DMSO), ethylene glycol (EG), propylene glycol (PG), methanol (MET), trehalose (TRE), and fetal bovine serum (FBS) were chosen [22]. All cryoprotectants used in the experiment, including DMSO, EG, PG, MET, TRE, and FBS, were manufactured by Xilong Scientific Co., Ltd., with a purity of 99% (analytical grade). Each cryoprotectant was mixed with Hank's solution to form the cryoprotective solution, achieving final concentrations of 5, 10, and 15% for DMSO, EG, PG, MET, TRE, and FBS (Table 2). Based on preliminary experiments that confirmed uniform sperm distribution and optimal cryopreservation performance, the prepared cryoprotectant solution was mixed with the sperm suspension at a consistent ratio of 10:1. Immediately after mixing, the solution was dispensed into 2 mL cryovials (each containing 1.5 mL) with all procedures performed on ice to form the experimental groups. Starting from the time of mixing, samples were equilibrated at 4 °C for 30 min and then transferred into a programmed freezer precooled to 4 °C. Each group was set up with three replicates.

Table 2. Experimental cryoprotectant solution data.

Group	Cryoprotectant Concentration (%)	Cryoprotectant Volume Added (μL)	Hank's Solution (μL)
Control group	/	/	1350
A	5% Dimethyl Sulfoxide (DMSO)	75	1275
	10% Dimethyl Sulfoxide (DMSO)	150	1200
	15% Dimethyl Sulfoxide (DMSO)	225	1125
B	5% Ethylene Glycol	59	1291
	10% Ethylene Glycol	118	1232
	15% Ethylene Glycol	177	1173
C	5% Propylene Glycol	73	1277
	10% Propylene Glycol	146	1204
	15% Propylene Glycol	219	1131
D	5% Methanol	43	1307
	10% Methanol	86	1264
	15% Methanol	129	1221
E	5% Trehalose	50	1300
	10% Trehalose	100	1250
	15% Trehalose	150	1200
F	5% Fetal Bovine Serum (FBS)	75	1275
	10% Fetal Bovine Serum (FBS)	150	1200
	15% Fetal Bovine Serum (FBS)	225	1125

2.6. Cooling and Thawing Procedures Configuration

In this experiment, a programmable freezing device (model CJ-L35, produced by Haier Biomedical Technology Co., Ltd., Chengdu, China) and an upright high-pressure liquid nitrogen container (model YDZ-125, also from Haier Biomedical Technology Co., Ltd.) were employed to cryopreserve the samples under controlled conditions. Through multiple preliminary tests, the optimal cooling procedure was identified: the temperature was lowered from 4 °C to −25 °C at 5 °C per min, held at −25 °C for 5 min, and then further reduced at 10 °C per min to −80 °C, where it was held for another 5 min. Prior to the loading of the sperm samples (both control and experimental), the freezing device was dried for 30 min and programmed to stabilize at 4 °C. Once the sperm samples reached the target temperature via the programmed protocol, they were swiftly transferred to the liquid nitrogen container.

Preliminary experiments demonstrated that a 38 °C water bath provided optimal thawing results. Therefore, both control and experimental samples were immersed in this 38 °C bath and gently swirled to achieve uniform thawing. After the ice in each cryovial had completely melted, the samples stood for approximately 10 s before being collected for subsequent analysis [23].

2.7. Assessment of Effect of Cryoprotectants on Motility and Kinematic Parameters of Sperm from M. nobilis

After thawing, sperm samples were mixed with sperm activation solution at a 1:10 ratio in a 15 mL conical tube. Subsequently, 5 μL of the mixture was placed in a sperm counting chamber for observation. A computer-assisted sperm analyzer was used to record sperm motility, curvilinear velocity, average path velocity, and straight-line velocity, with each group tested in triplicate. The CASA system settings were as follows: a maximum capture area of 80 im^2, a frame rate of 300 f/s, and a sample drift parameter of 4 pm/s, with each field capturing between 300 and 400 sperm. These settings ensured the accurate and consistent measurement of sperm motion parameters for reliable subsequent analysis.

Subsequently, the sperm obtained from these three individuals were pooled to form a homogeneous composite sample as the singular experimental unit for all subsequent assays.

2.8. Assessment of Effect of Cryoprotectants on Morphology and Ultrastructure of Sperm

Based on the effects of different cryoprotectants on *M. nobilis* sperm, fresh sperm samples, as well as samples from the best and worst experimental conditions, were selected. A 1 mL aliquot of each sample was fixed with a solution of 4% paraformaldehyde and 1% glutaraldehyde (prepared using 50% glutaraldehyde 10 mL, 20 g paraformaldehyde, and 3.029 g Tris buffer, dissolved in 500 mL filtered seawater). The fixed samples were subjected to dehydration in a graded ethanol series, followed by the addition of isoamyl acetate. After critical point drying using a LEICA EM CPD 300 critical point dryer (produced by Leica Microsystems, Beijing, China) and coating with gold using an IB-5uj ion sputter coater, the morphology of the sperm was imaged and observed with a Phenom ProX G6 scanning electron microscope (manufactured by Phenom Scientific Instruments, Shanghai, China).

For ultrastructural observation, fresh sperm samples and those from the most and least favorable experimental conditions were initially rinsed with precooled 0.1 mol/L phosphate buffer at 4 °C. They were then immersed in 2.5% glutaraldehyde fixative under dark conditions at 4 °C for 4 h. After this fixative was discarded, the samples were placed in fresh 2.5% glutaraldehyde solution for overnight preservation at 4 °C. Next, they were rinsed three times in 0.1 mol/L phosphate buffer (15 min each), followed by fixation in 1% osmium tetroxide for 2 h. The osmium tetroxide was then removed, and the samples were again rinsed three times in 0.1 mol/L phosphate buffer (15 min each). Dehydration was performed using a graded ethanol series (30, 50, 70, 80, 90, and 100%) and ethanol–acetone mixtures (3:1, 1:1, and 1:3), and then washed twice in pure acetone (10 min each). After this, the samples were embedded in resin and subjected to high-temperature polymerization once the resin had fully solidified. Ultrathin sections (70–80 nm) were prepared using an ultramicrotome (UC Enuity ultramicrotome, Leica Microsystems, Beijing, China) and double-stained with 5% uranyl acetate and lead citrate. Finally, sperm ultrastructure images were captured using a Spectra Ultra S/TEM transmission electron microscope (Thermo Fisher Scientific).

Subsequently, the sperm obtained from these three individuals were pooled to form a homogeneous composite sample as the singular experimental unit for all subsequent assays.

2.9. Sperm Quality Assessment

After the recovery procedure, *M. nobilis* sperm were activated using the designated activation medium, and detailed observations were documented. Each sample was then analyzed using a sperm quality analyzer, capturing 400–600 sperm per measurement. Sperm vitality was determined by examining six randomly selected fields of view. Key parameters assessed for sperm quality included sperm vitality (combined percentage of progressive and non-progressive motility), curvilinear velocity (VCL), average path velocity (VAP), and straight-line velocity (VSL) [24].

Subsequently, the sperm obtained from these three individuals were pooled to form a homogeneous composite sample as the singular experimental unit for all subsequent assays.

2.10. Data Analysis

Sperm parameter data collected via the Computer-Assisted Sperm Analysis (CASA) system were analyzed using SPSS 27.0. Prior to the one-way ANOVA, arcsine transformation was applied to ensure adherence to normal distribution assumptions. Distribution tests were also performed to verify normality and the homogeneity of variances, confirming the appropriateness of the variance analysis. All necessary statistical assumptions were met, affirming the validity of the ANOVA results. Significance was set at $p < 0.05$, with $p < 0.01$ indicating a highly significant difference, and $p < 0.001$ denoting an extremely significant difference. Results are expressed as mean ± standard deviation (mean ± S.D). All graphs were created using Origin 2021.

3. Results

3.1. Effects of Cryoprotectants on Sperm Motility in M. nobilis

The motility of *M. nobilis* sperm after cryopreservation and thawing was assessed (Figure 1). In the control group without cryoprotectants, sperm motility was 13.76 ± 1.04%. Among the experimental groups with DMSO added as a cryoprotectant at concentrations ranging from 5 to 15% (Figure 1A), sperm motility was the highest, with 5% DMSO at 51.68 ± 2.36%, 10% DMSO at 69.10 ± 2.37%, and 15% DMSO at 54.59 ± 2.54%. Sperm motility in all DMSO groups was significantly higher than in the control group ($p < 0.001$) and superior to the other experimental groups. Notably, 10% DMSO achieved the highest motility, minimally affecting sperm compared to fresh sperm motility at 81.32 ± 2.48%.

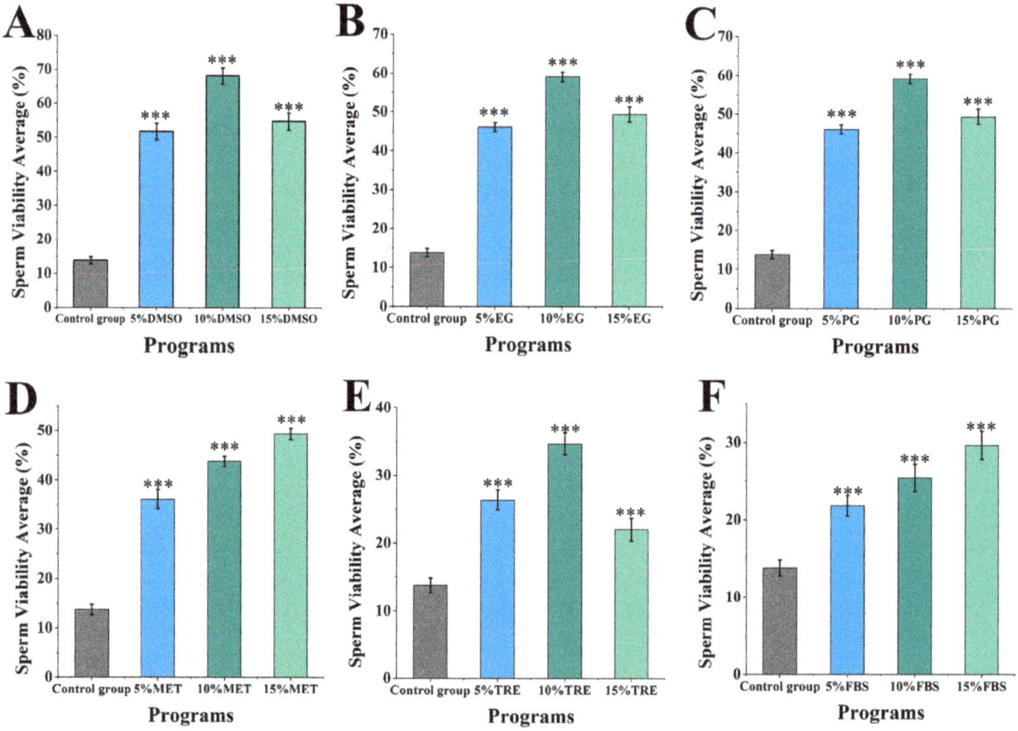

Figure 1. Sperm quality in different cryoprotectants at different concentrations. (**A**) Dimethyl sulfoxide (DMSO) at concentrations of 5%, 10%, and 15%; (**B**) Ethylene glycol (EG) at concentrations of 5%, 10%, and 15%; (**C**) Propylene glycol (PG) at concentrations of 5%, 10%, and 15%; (**D**) Methanol (MET) at concentrations of 5%, 10%, and 15%; (**E**) Trehalose (TRE) at concentrations of 5%, 10%, and 15%; (**F**) Fetal bovine serum (FBS) at concentrations of 5%, 10%, and 15% (*** $p < 0.001$).

In the groups with MET and FBS as cryoprotectants (Figure 1D,F), sperm motility displayed a positive correlation with increasing concentration. In the MET group, motility was 36.19 ± 1.95% for 5% MET, 43.81 ± 1.02% for 10% MET, and 49.38 ± 1.14% for 15% MET. As MET concentration increased, sperm motility significantly rose, with the 15% MET group displaying motility notably higher than the 5 and 10% groups ($p < 0.001$). In the FBS group, motility was 25.43 ± 1.33% for 5% FBS, 25.43 ± 1.76% for 10% FBS, and 29.65 ± 1.82% for 15% FBS. At 15% FBS, sperm motility was significantly higher than at 5 and 10% FBS ($p < 0.001$). Although MET and FBS provided varying protective effects, both demonstrated increased motility with rising concentrations.

In the other cryoprotectant groups (ethylene glycol, propylene glycol, and trehalose), sperm motility at all concentrations was also significantly higher than in the control group ($p < 0.001$) (Figure 1B,C,E), though cryopreservation effectiveness across these concentrations did not match that of DMSO. Overall, the effects of different cryoprotectants on sperm motility varied, with 10% DMSO being the most effective and minimally impacting sperm motility.

3.2. Impact of Cryoprotectants on Curvilinear Velocity of M. nobilis Sperm

The curvilinear velocity of *M. nobilis* sperm after cryopreservation and thawing was assessed (Figure 2). In the control group without cryoprotectants, the curvilinear velocity was 33.53 ± 1.63 μm/s. When DMSO was introduced as a cryoprotectant, the sperm's curvilinear velocity increased significantly, outperforming all other experimental groups. In particular, sperm treated with 10% DMSO exhibited the highest curvilinear velocity at 84.94 ± 1.64 μm/s, significantly higher than that of the control group ($p < 0.001$). This rate closely approximates the motility of fresh, untreated sperm at 103.64 ± 2.85 μm/s, indicating minimal impact on curvilinear velocity. Other DMSO concentrations also showed effective preservation, with 5 and 15% DMSO resulting in velocities of 70.13 ± 1.57 and 76.81 ± 1.82 μm/s, respectively, both significantly higher than those of the control group ($p < 0.001$).

In the experimental groups with MET and FBS as cryoprotectants (Figure 2D,F), increased concentrations of MET and FBS were associated with proportional increases in curvilinear velocity. In the MET group, 5% MET achieved a velocity of 48.9 ± 2.05 μm/s, 10% MET reached 53.26 ± 1.81 μm/s, and 15% MET reached 61.52 ± 1.55 μm/s. As MET concentration increased, curvilinear velocity improved significantly, with the 15% MET group showing a significantly higher velocity than the 5 and 10% groups ($p < 0.05$). In the FBS group, the curvilinear velocities were 38.72 ± 1.52 μm/s for 5% FBS, 42.98 ± 1.68 μm/s for 10% FBS, and 49.37 ± 1.79 μm/s for 15% FBS, with the 15% FBS group significantly outperforming the 5 and 10% groups ($p < 0.05$). While the protective effects of MET and FBS varied, both demonstrated an upward trend in velocity with increasing concentration.

In the other cryoprotectant groups (ethylene glycol, propylene glycol, and trehalose), sperm curvilinear velocity at all concentrations was also significantly higher than in the control group ($p < 0.001$) (Figure 2B,C,E), though none of these concentrations matched the effectiveness of DMSO. Overall, the impact of different cryoprotectants on sperm curvilinear velocity varied, with 10% DMSO exhibiting the most notable effect and minimal impact on velocity.

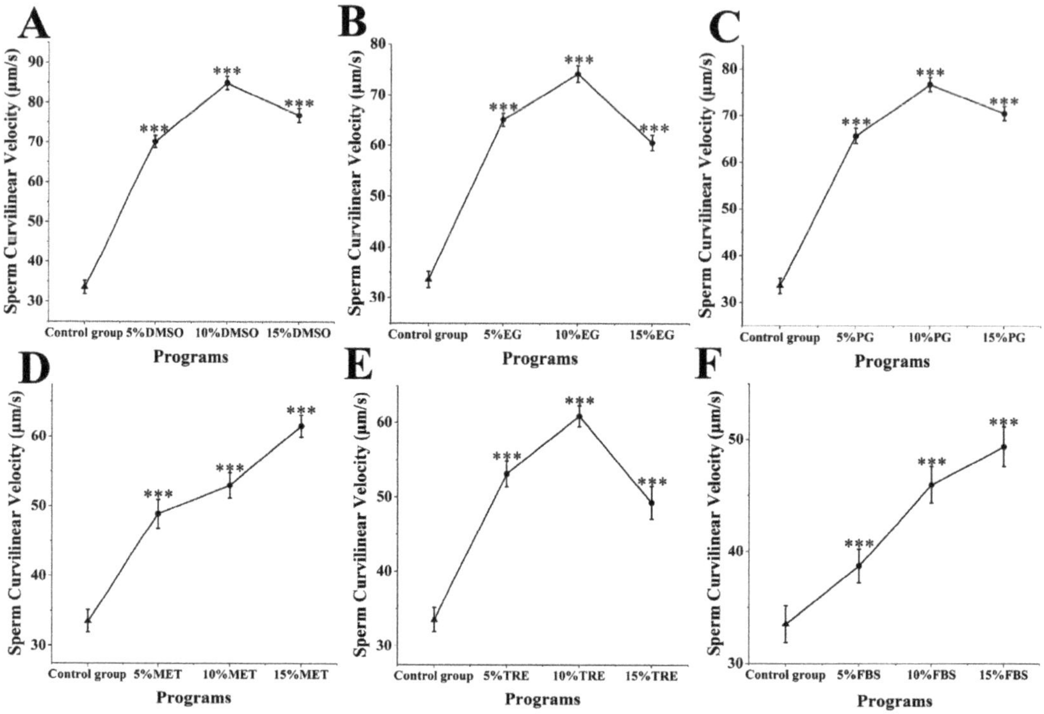

Figure 2. Impact of cryoprotectants on curvilinear velocity of sperm from *M. nobilis*. (**A**) Dimethyl sulfoxide (DMSO) at concentrations of 5%, 10%, and 15%; (**B**) Ethylene glycol (EG) at concentrations of 5%, 10%, and 15%; (**C**) Propylene glycol (PG) at concentrations of 5%, 10%, and 15%; (**D**) Methanol (MET) at concentrations of 5%, 10%, and 15%; (**E**) Trehalose (TRE) at concentrations of 5%, 10%, and 15%; (**F**) Fetal bovine serum (FBS) at concentrations of 5%, 10%, and 15% (*** $p < 0.001$).

3.3. *Impact of Cryoprotectants on Average Path Velocity of* M. nobilis *Sperm*

The average path velocity of *M. nobilis* sperm was measured following cryopreservation and thawing (Figure 3). In the control group, which lacked cryoprotectants, the sperm's average path velocity was 21.68 ± 1.76 μm/s. When DMSO was added as a cryoprotectant (Figure 3A), the average path velocity of the sperm increased markedly, outperforming all other groups. Notably, 10% DMSO resulted in the highest average path velocity at 68.77 ± 2.05 μm/s, significantly surpassing the control group ($p < 0.001$). This velocity was close to that of fresh, untreated sperm at 81.35 ± 2.36 μm/s, indicating minimal reduction in velocity. Other concentrations of DMSO also provided strong protection, with velocities of 56.53 ± 1.87 μm/s for 5% DMSO and 59.36 ± 1.75 μm/s for 15% DMSO, both significantly higher than those of the control group ($p < 0.001$).

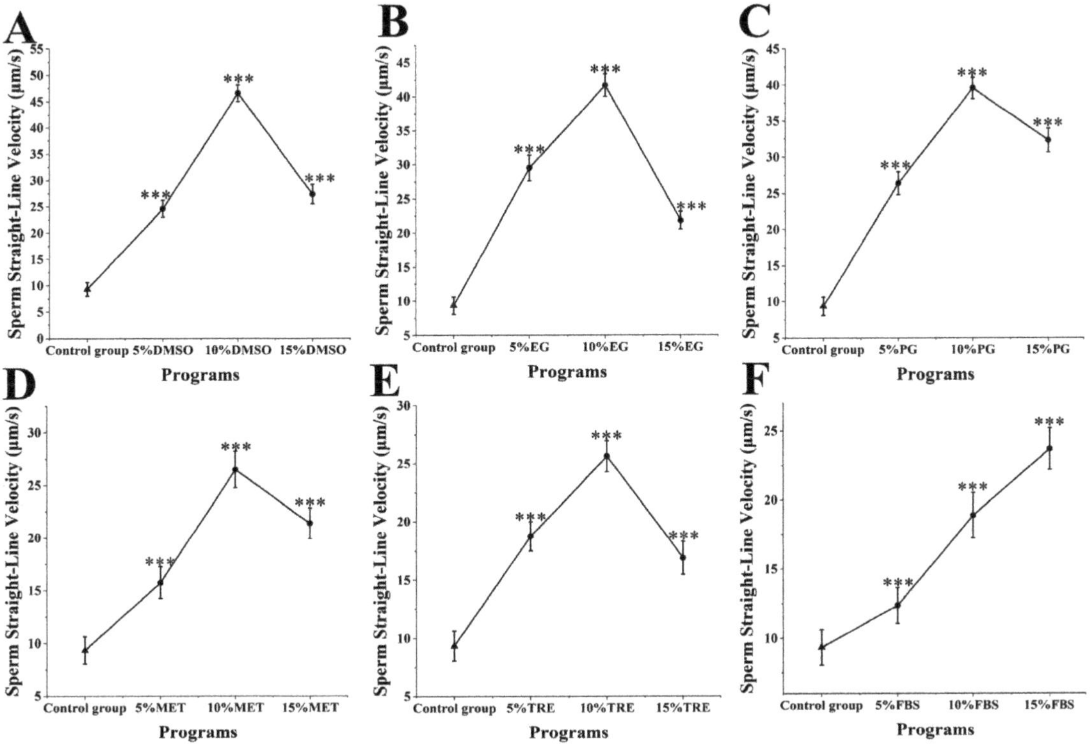

Figure 3. Impact of cryoprotectants on average path velocity of sperm from *M. nobilis*. (**A**) Dimethyl sulfoxide (DMSO) at concentrations of 5%, 10%, and 15%; (**B**) Ethylene glycol (EG) at concentrations of 5%, 10%, and 15%; (**C**) Propylene glycol (PG) at concentrations of 5%, 10%, and 15%; (**D**) Methanol (MET) at concentrations of 5%, 10%, and 15%; (**E**) Trehalose (TRE) at concentrations of 5%, 10%, and 15%; (**F**) Fetal bovine serum (FBS) at concentrations of 5%, 10%, and 15% (*** $p < 0.001$).

For the MET and FBS groups (Figure 3D,F), the average path velocity showed a positive relationship with increasing concentrations of these cryoprotectants. In the MET group, the average velocities were 34.59 ± 1.88 μm/s for 5% MET, 46.63 ± 1.75 μm/s for 10% MET, and 54.01 ± 1.63 μm/s for 15% MET. As methanol concentration increased, the sperm's average path velocity significantly improved, with the 15% MET group showing a substantially higher velocity than the 5 and 10% groups ($p < 0.05$). In the FBS group, the average path velocities were 26.64 ± 1.92 μm/s for 5% FBS, 32.71 ± 1.57 μm/s for 10% FBS, and 39.43 ± 1.75 μm/s for 15% FBS, with 15% FBS notably exceeding the lower concentrations ($p < 0.05$). Although MET and FBS showed different protective effects, both followed an upward trend in velocity as concentration increased.

In the remaining cryoprotectant groups (ethylene glycol, propylene glycol, and trehalose), the average path velocities for all concentrations were significantly greater than in the control group ($p < 0.001$) (Figure 3B,C,E). However, these cryoprotectants were generally less effective than DMSO. Overall, the influence of different cryoprotectants on the sperm's average path velocity varied, with 10% DMSO providing the most significant protective effect, resulting in the smallest reduction in sperm velocity.

3.4. Impact of Cryoprotectants on Straight-Line Velocity of *M. nobilis* Sperm

The straight-line velocity of *M. nobilis* sperm post-cryopreservation and thawing was analyzed (Figure 4). In the control group, which did not include any cryoprotectants, sperm straight-line velocity was 9.35 ± 1.28 μm/s. When DMSO was applied as a cryoprotectant (Figure 4A), the sperm's straight-line velocity increased significantly compared to all other groups. Specifically, treatment with 10% DMSO yielded the highest velocity at 46.57 ± 1.57 μm/s, substantially surpassing the control group ($p < 0.001$). This rate was notably close to that of fresh sperm at 58.27 ± 1.58 μm/s, showing minimal impact on straight-line velocity. Other DMSO concentrations also provided effective protection, with velocities of 24.63 ± 1.62 μm/s for 5% DMSO and 27.36 ± 1.78 μm/s for 15% DMSO, both significantly higher than those of the control group ($p < 0.001$).

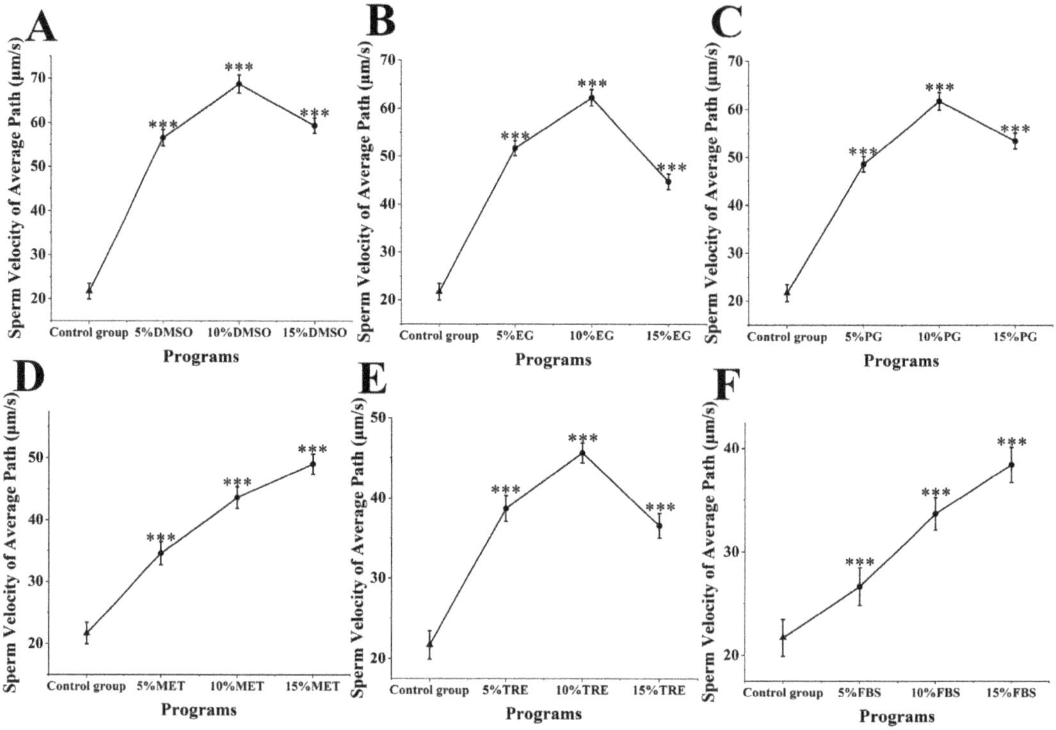

Figure 4. Impact of cryoprotectants on straight-line velocity of Sperm from *M. nobilis*. (**A**) Dimethyl sulfoxide (DMSO) at concentrations of 5%, 10%, and 15%; (**B**) Ethylene glycol (EG) at concentrations of 5%, 10%, and 15%; (**C**) Propylene glycol (PG) at concentrations of 5%, 10%, and 15%; (**D**) Methanol (MET) at concentrations of 5%, 10%, and 15%; (**E**) Trehalose (TRE) at concentrations of 5%, 10%, and 15%; (**F**) Fetal bovine serum (FBS) at concentrations of 5%, 10%, and 15% (*** $p < 0.001$).

In groups using MET and FBS as cryoprotectants (Figure 4D,F), a positive correlation was observed between cryoprotectant concentration and straight-line velocity. In the MET group, 5% MET yielded a velocity of 15.75 ± 1.52 μm/s, 10% MET yielded 27.48 ± 1.92 μm/s, and 15% MET reached 21.35 ± 1.78 μm/s. Increasing the MET concentration significantly raised sperm straight-line velocity, with the 15% MET group surpassing both the 5% and 10% groups ($p < 0.05$). In the FBS group, the velocities were 12.36 ± 1.28 μm/s for 5% FBS, 17.87 ± 1.65 μm/s for 10% FBS, and 23.68 ± 1.51 μm/s for 15% FBS, with the 15% concentration showing a significant improvement over the lower concentrations ($p < 0.05$). Despite differing protective effects, both MET and FBS showed a trend of enhanced straight-line velocity with increasing concentrations.

In other experimental groups using ethylene glycol, propylene glycol, and trehalose, sperm straight-line velocity at all concentrations was also significantly higher than in the control group ($p < 0.001$) (Figure 4B,C,E), though the protective effects across concentrations did not reach the levels achieved by DMSO. Overall, the influence of various cryoprotectants on straight-line velocity varied, with 10% DMSO providing the most prominent protective effect and having the least impact on straight-line velocity.

3.5. Effects of Cryoprotectants on Sperm Morphology and Ultrastructure in M. nobilis

Based on its superior performance—yielding a sperm motility of 69.10 ± 2.37%, curvilinear velocity of 84.94 ± 1.64 μm/s, average path velocity of 68.77 ± 2.05 μm/s, and straight-line velocity of 46.57 ± 1.57 μm/s—compared to other cryoprotectants, 10% DMSO proved to be the best cryoprotectant. Therefore, we selected samples from fresh sperm, sperm cryopreserved with 10% DMSO, and a control group without cryoprotectants to examine sperm morphology and ultrastructure.

The *M. nobilis* sperm are flagellated (Figure 5A) and measure approximately 60–70 μm in length, consisting of three main parts: the head, midpiece, and tail. The head, which is about 4.2–5.0 μm long and conical in shape, contains the acrosome and nucleus. The acrosome, approximately 0.7 μm long and conical, is situated at the very tip of the head. The nucleus is cylindrical, narrower at the top and wider at the base, with a length of around 3.6 μm. The midpiece is short, approximately 0.9–1.0 μm in length, while the tail is elongated, consisting of an axoneme and plasma membrane, with the axoneme extending from the distal centriole. After cryopreservation with 10% DMSO, partial fragmentation of the flagellum was observed in *M. nobilis* sperm (Figure 5B). In contrast, in the control group without cryoprotectants, the sperm flagella exhibited complete breakage (Figure 5C).

Ultrastructural examination revealed that fresh *M. nobilis* sperm were abundant, exhibiting no detectable swelling or structural damage (Figure 6A). The cell membranes were intact, continuous, and sharply defined, with densely packed cytoplasm and smoothly outlined organelles. The nuclei maintained a distinct conical shape, characterized by uniformly dispersed chromatin. The mitochondria were well developed, showing a clear, structurally organized matrix and intact, neatly arranged cristae. Moreover, the acrosome structure was fully preserved and clearly demarcated, and the microtubules within the tail region were prominently visible and structurally sound, indicating overall excellent sperm ultrastructure.

In contrast, sperm samples treated with 10% DMSO (Figure 6B) maintained relatively high sperm counts but showed moderate ultrastructural damage. Although the majority of the cell membranes remained intact, certain regions appeared slightly blurred or compromised, accompanied by a visible decrease in cytoplasmic density. The organelles displayed evident swelling, particularly the mitochondria, which showed signs of membrane disruption and exhibited an irregular matrix structure with a slight deformation or loss of cristae (indicated by red arrows). The nucleus generally retained its conical form with evenly dispersed chromatin, indicating minimal nuclear damage. The acrosome was mostly intact, although microtubule structures appeared somewhat indistinct or obscured (marked by yellow arrows), suggesting mild impairment of the sperm tail structure.

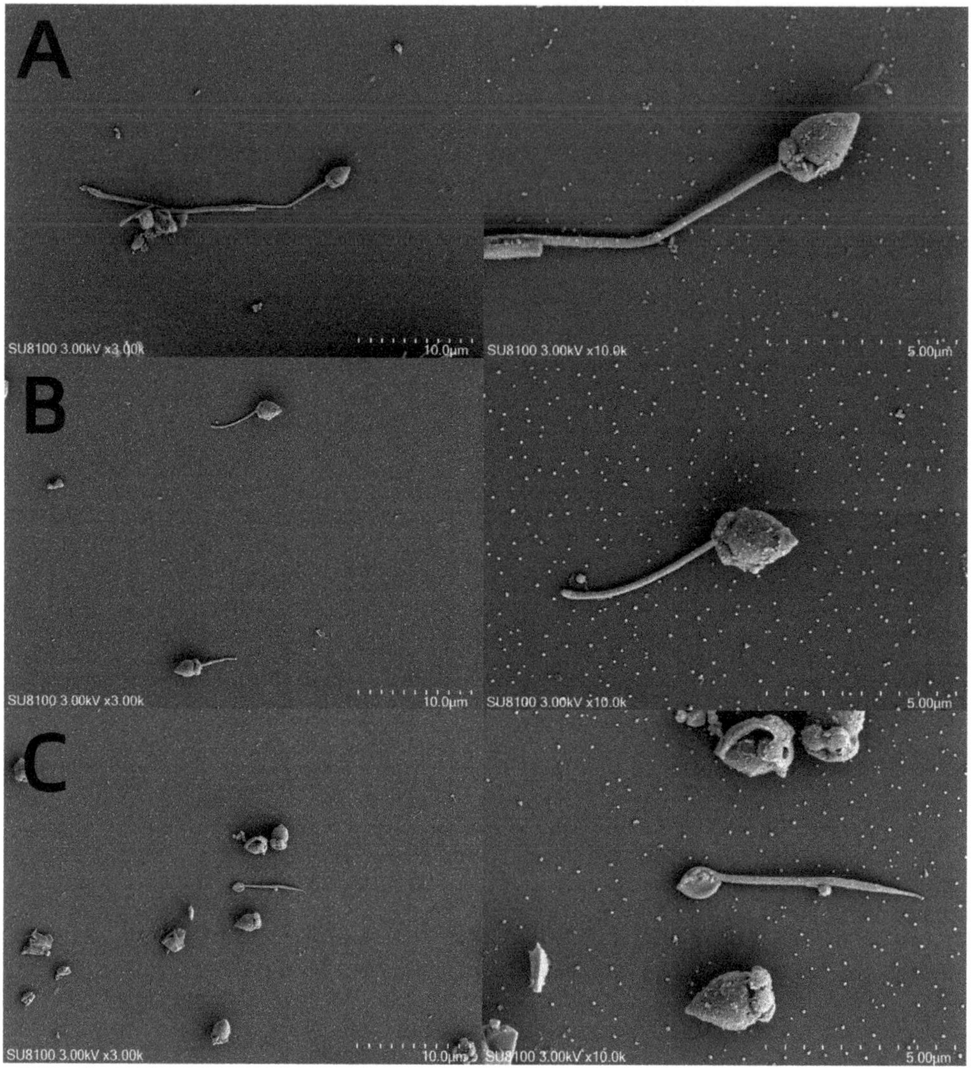

Figure 5. Scanning electron micrographs of sperm subjected to different cryoprotectants. (**A**) Fresh sperm, (**B**) cryopreserved sperm, and (**C**) control group without cryoprotectant. Note: variation in structure of head and tail in all tested samples.

In the control group lacking cryoprotectants (Figure 6C), the number of sperm was reduced, and they exhibited severe swelling. The cell membranes were compromised and partially dissolved, while the cytoplasm displayed extensive low-electron-density areas, indicative of pronounced organelle swelling. Although the nuclei appeared irregular, the chromatin was uniformly distributed. Mitochondria showed marked swelling and vacuolization, with blurred or dissolved membranes, widespread matrix degradation, and a significant loss or disruption of cristae (red arrows). Microtubules were barely detectable.

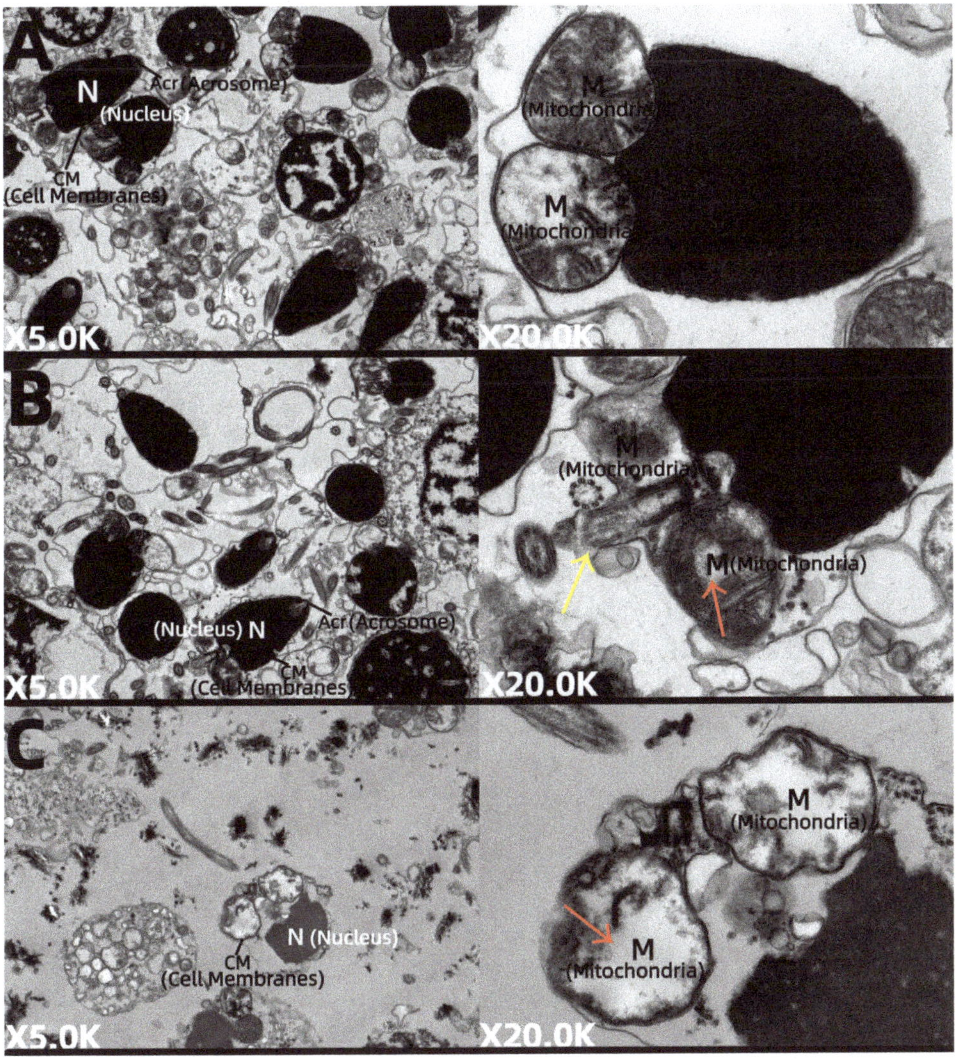

Figure 6. Transmission electron micrographs of sperm subjected to different cryoprotectants. (**A**) Fresh sperm, (**B**) cryopreserved sperm, and (**C**) control group without cryoprotectant. Note: variation in cellular organelles in different treatments. Nucleus (N); acrosome (Ac); mitochondria (M).

4. Discussion

Cryoprotective agents are compounds that, during ultralow-temperature freezing, help narrow dangerous temperature ranges and reduce freezing damage, thereby protecting sperm. However, they can cause cellular damage, and different animals respond to cryoprotectants with varying mechanisms, leading to differences in effectiveness [25]. Based on their mode of action, cryoprotectants are classified as permeable or non-permeable. Among the factors influencing the efficacy of sperm cryopreservation, selecting the appropriate type and concentration of cryoprotectant is critical, as it significantly affects preservation outcomes [26]. In this study, six cryoprotectants were used, both permeable (DMSO, EG, PG and MET) and non-permeable (TRE and FBS). Permeable agents are generally small, neutral molecules that easily bind water molecules in solution, promoting hydration and increasing solution viscosity, which reduces ice crystal formation and protects cells [27]. Adding permeable cryoprotectants during freezing also dilutes the solute concentration, reducing cellular salt intake as the cryoprotectant replaces it [28]. However, due to their inherent toxicity, permeable cryoprotectants can cause harm at higher concentrations, as toxic effects may outweigh cryoprotective benefits, necessitating a balance to optimize their protective effects [29]. The results from this study indicate that DMSO provided the best cryoprotection for *M. nobilis* sperm, preserving motility, sperm velocity, and morphology. This effectiveness likely results from DMSO's ability to quickly penetrate sperm cell membranes and interact with phospholipids [30], aligning with findings in prior studies on Australian flat oyster (*Ostrea angasi*) sperm cryopreservation [31]. Research has shown that using 10% DMSO as a cryoprotectant in Eastern oyster (*Crassostrea virginica*) sperm cryopreservation resulted in a 70% post-thaw survival rate [32]. In another study, using sperm motility and fertilization rate as parameters, the researchers found that sperm cryopreserved with a 1:3 mixture of 10% DMSO and seawater achieved over 40% motility and a 20% fertilization rate, with a portion of fertilized eggs successfully developing to the D-larval stage [33]. Additionally, a study on Chinese pearl oyster (*Pinctada martensii*) sperm reported post-thaw survival rates exceeding 60% and fertilization rates up to 80% when cryopreserved with a 10% DMSO–seawater solution [34]. Thus, species-specific physiological characteristics and biological specificity should be considered in cryopreservation studies.

Non-permeable cryoprotectants create an osmotic pressure gradient around cells, facilitating water efflux and reducing intracellular ice formation during freezing. These cryoprotectants dissolve in water but do not penetrate cells, enabling the solution to remain supercooled and lowering solute concentration at specific temperatures to protect cells [35]. In this study, TRE and FBS served as non-permeable cryoprotectants, though their effects on *M. nobilis* sperm motility and swimming speed post-freezing were suboptimal. This is likely because TRE and FBS cannot penetrate cell membranes, thus failing to prevent intracellular ice crystal formation, which is essential for cell survival [36]. Furthermore, they do not alleviate the effects of extreme extracellular environmental changes, such as osmotic pressure shifts and ice crystal damage, leading to cell structure and function loss [37]. Research indicates that adding 0.45 mol/L trehalose to Japanese pearl oyster (*Pinctada fucata martensii*) sperm cryoprotectant significantly improved motility and fertilization rates post-thaw compared to untreated groups [38]. In green-lipped mussel (*Perna canaliculus*) sperm cryopreservation, a cryoprotectant mixture of 12% DMSO and 0.2 mol/L trehalose yielded effective results [39]. Another study on Japanese pearl oyster sperm cryopreservation compared 10% methanol, dimethylformamide, and 10% DMSO with a diluent of 80% seawater and 20% bovine serum, finding that methanol provided superior post-thaw motility [40]. Using mixed cryoprotectants from different classes may provide enhanced protection, though the specific mechanisms require further exploration [41].

The optimal concentration of cryoprotectants is closely related to species specificity. Cells in different species vary significantly in membrane permeability, sensitivity to freezing stress, metabolic activity, and solute and water distribution inside and outside cells [42], all of which determine species-specific cryoprotectant concentration requirements. In this experiment, a 10% concentration of each cryoprotectant (DMSO, EG, PG, TRE, and FBS) provided superior motility and velocity for *M. nobilis* sperm compared to 5 and 15% concentrations. A 15% concentration, though increasing cryoprotection, led to higher cellular toxicity, reducing sperm motility and sperm velocity. This may be due to the high concentration of DMSO causing a drastic increase in extracellular osmotic pressure and an excessive amount of DMSO entering the cells, which disrupts the integrity of the cell membrane, leading to cell dehydration and intracellular imbalance [43]. Studies have shown that during cryopreservation, cobia (*Rachycentron canadum*) sperm exhibited a 16%

greater loss of motility under 15% DMSO compared to a 10% concentration [44]. Consequently, protective effects are often offset by toxic effects, resulting in reduced sperm activity and viability. Meanwhile, 5% was too low to offer effective protection. Notably, sperm cryopreserved with MET and FBS showed a positive correlation between cryoprotectant concentration and cryoprotective efficacy, likely due to MET's low cytotoxicity and freezing point, which minimizes ice formation and effectively reduces sample freezing temperature [45]. MET's rapid action and easy removal from cells minimize long-term chemical effects, while its dehydrating properties help maintain cell structure [46]. The protein molecules in FBS stabilize the extracellular environment, reducing osmotic damage during freezing and protecting cell membranes through specific interactions with cell surfaces [47]. FBS also provides nutritional support and promotes cell adhesion [48]. Consequently, MET and FBS are more suitable for preserving high bioactivity and cell integrity in sensitive biological samples at high concentrations, whereas other cryoprotectants may be less effective in providing added protection at elevated concentrations [49]. Studies show that 10% methanol as a cryoprotectant resulted in the highest post-thaw motility and longest swimming duration for zebrafish (*Danio rerio*) sperm [50]. Additionally, bovine serum protein outperformed trehalose in cryopreserving sea bream (*Sparus aurata*) sperm, as trehalose's protective mechanism relies on forming an ordered "glassy" state to prevent extracellular ice crystal formation, physically stabilizing the environment around cells [51]. This mechanism depends on trehalose concentration and freeze–thaw control; otherwise, it can reduce sperm motility [52]. For Tambaqui (*Colossoma macropomum*) sperm, trehalose had the greatest impact on survival rate, supporting these findings [53]. Further research on the effects of cryoprotectant concentration on sperm cryopreservation should incorporate species-specific characteristics and biochemical mechanisms.

In the morphological and ultrastructural observation of cryopreserved *M. nobilis* sperm, we found that cryopreservation induces similar types of damage across different species, specifically affecting the sperm membrane, acrosome, mitochondria, and flagellum. These damages disrupt membrane integrity and functional proteins [54]. In this experiment, we focused on sperm samples under fresh, optimal (10% DMSO), and control (no cryoprotectant) conditions. Observations showed varying degrees of morphological damage in cryopreserved sperm, primarily concentrated in the flagellum, a critical structure for sperm motility. Flagellar integrity directly impacts sperm motility and viability [55], and partial flagellar breakage likely results from membrane stress responses during thawing, negatively affecting sperm motility to some extent [56]. Under 10% DMSO, flagellar breakage was minimal, preserving some motility. In contrast, complete flagellar breakage in the control group led to significant declines in sperm motility and may have impaired fertilization potential, aligning with this experiment's findings on sperm motility and sperm velocity. Research on red abalone (*Haliotis rufescens*) sperm cryopreservation found that cryopreserved sperm exhibited damage to the flagellum or plasma membrane, leading to significant declines in motility and fertilization rates compared to fresh sperm [57]. Additionally, in European eel (*Anguilla anguilla*) sperm cryopreservation, non-motile sperm could still fertilize eggs and induce the acrosome reaction, suggesting that sperm with flagellar detachment or membrane damage may retain fertilization potential [58]. In this study, ultrastructural examinations revealed multiple types of damage, including varying degrees of membrane swelling, enlarged intermembrane spaces, ruptured or absent acrosomes, the disintegration of mitochondria, distorted cristae, and the separation of the double membrane. Swelling, fractures, and even detachment of flagella were also commonly observed. During freezing and thawing, ice crystals may mechanically damage cell membranes to different extents [59], while low temperatures induce membrane phase transitions that reduce elasticity and increase the likelihood of rupture. This leads to membrane protein loss, protein conformational changes, and cytoplasmic alterations [60].

Cryopreservation-induced damage to the sperm tail disrupts the axonemal structure, which diminishes motility and ultimately impairs sperm function [61]. During the freezing process, cryoprotectants that penetrate cells raise intracellular osmotic pressure above extracellular levels, causing extracellular water influx, further swelling, and membrane rupture, especially in the sperm tail [62]. Studies on grouper (*Epinephelus coioides*) sperm cryopreservation showed that damage to sperm membranes and a loss of or alteration in membrane proteins disrupted the interaction required for sperm–egg recognition, thereby blocking sperm penetration of the egg [63]. Further research is needed to elucidate the specific mechanisms by which cryoprotectants impact sperm integrity.

5. Conclusions

This study demonstrates that the type and concentration of cryoprotectant are crucial for the cryopreservation efficacy of *M. nobilis* sperm. Permeable cryoprotectants, such as DMSO, significantly enhance the motility and morphological preservation of *M. nobilis* sperm at optimal concentrations, with 10% DMSO showing the best results. Meanwhile, non-permeable cryoprotectants like FBS also exhibit stabilizing effects on the extracellular environment at higher concentrations. As concentration increases, both MET and FBS provide progressively enhanced protection, further reducing damage to the sperm. Additionally, damage incurred during the freezing process directly affects the motility of *M. nobilis* sperm, with 10% DMSO preserving high integrity and motility. These findings underscore the importance of optimizing cryoprotectant selection and concentration, providing key insights for the germplasm preservation of marine species.

Author Contributions: Conceptualization, M.L., Z.F. and Z.M.; methodology, M.L.; software, M.L.; validation, M.L., Z.F. and Z.M.; formal analysis, M.L.; investigation, M.L. and Z.F.; resources, G.Y. and Z.M.; data curation, M.L.; writing—original draft preparation, M.L.; writing—review and editing, M.L., G.Y. and Z.M.; visualization, M.L.; supervision, G.Y. and Z.M.; project administration, G.Y. and Z.M.; funding acquisition, Z.M. All authors have read and agreed to the published version of the manuscript.

Funding: This work was supported by the Central Public-interest Scientific Institution Basal Research Fund, CAFS [grant number 2023TD58], the earmarked fund for CARS [grant number CARS-49], and the Financial Fund of Ministry of Agriculture and Rural affairs of China [grant number NHYYSWZZZYKZX2020].

Institutional Review Board Statement: The noble scallop *Mimachlamys nobilis* is not a vertebrate nor a regulated invertebrate, so no ethical conduct approval is required.

Data Availability Statement: The raw data supporting the conclusions of this article will be made available by the authors on request.

Conflicts of Interest: The authors declare no conflicts of interest. The funders had no role in the design of the study; in the collection, analyses, or interpretation of data; in the writing of the manuscript; or in the decision to publish the results.

Abbreviations

The following abbreviations are used in this manuscript:

mm	Millimeter
μm	Micron
L	Liter
mL	Milliliter
μL	Microliter
min	Minute
s	Second
g	Gram
°C	Centigrade
DMSO	Dimethyl Sulfoxide
EG	Ethylene Glycol
PG	Propylene Glycol
MeOH	Methanol
TRE	Trehalose
FBS	Fetal Bovine Serum
CASA	Computer-Assisted Sperm Analysis
Hanks's	Hank's Balanced Salt Solution
SEM	Scanning Electron Microscopy
TEM	Transmission Electron Microscope

References

1. Xu, C.; Wu, X.; Qiu, J.; Ye, J.; Lin, Q.; Deng, J.; Zeng, Y.; Wang, W.; Zhang, H.; Zheng, H. Genome-wide identification of gap junction gene family and their expression profiles under low temperature stress in noble scallop *Chlamys nobilis*. *Comp. Biochem. Physiol. Part D Genom. Proteom.* **2024**, *52*, 101310. [CrossRef] [PubMed]
2. Li, M.; Wu, J.; Yang, R.; Fu, Z.; Yu, G.; Ma, Z. Effects of ammonia concentration on sperm vitality, motility rates, and morphology in three marine bivalve species: A comparative study of the noble scallop *Mimachlamys nobilis*, Chinese pearl oyster *Pinctada fucata martensii*, and small rock oyster *Saccostrea mordax*. *Biology* **2024**, *13*, 589. [CrossRef]
3. Zhang, C.; Sun, Y.; Wen, J.; Xu, B.; Zhu, W.; Zhang, H.; Liu, X.; Lichu, L.; Zheng, H. Effects of chronic cold stress on tissue structure, antioxidant response, and key gene expression in the warm-water bivalve *Chlamys nobilis*. *Comp. Biochem. Physiol. Part D Genom. Proteom.* **2024**, *50*, 101225. [CrossRef] [PubMed]
4. Zhu, W.; Zhang, C.; Tan, K.; Wang, B.; Huang, R.; Wen, J.; Xu, B.; Liu, X.; Lichu, L.; Zheng, H. Variation of lipids and fatty acids in noble scallop *Chlamys nobilis* under low temperature stress. *Aquaculture* **2022**, *554*, 738121. [CrossRef]
5. Zheng, X.; Lin, S.; Yang, R.; Qin, J.; Wang, A.; Gu, Z.; Ma, Z. Sperm cryopreservation of the noble scallop *Chlamys nobilis* by a programmable freezing method: Effect of cryoprotectant. *Aquac. Res.* **2019**, *50*, 1678–1686. [CrossRef]
6. Barbas, J.P.; Mascarenhas, R.D. Cryopreservation of domestic animal sperm cells. *Cell Tissue Bank.* **2009**, *10*, 49–62. [CrossRef]
7. Tambovsky, M.A.; Aimaletdinov, A.M.; Zakirova, E.Y. Current trends in the application of stem cells and their derivatives in animal sperm cryopreservation. *Biochem. (Mosc.) Suppl. Ser. A Membr. Cell Biol.* **2023**, *17*, 243–248. [CrossRef]
8. Chelewani, A.P.; Takahashi, E.; Nishimura, T.; Fujimoto, T. Optimizing the post-thaw quality of cryopreserved masu salmon (*Oncorhynchus masou*) sperm: Evaluating the effects of antioxidant-supplemented extender. *Aquaculture* **2024**, *593*, 741332. [CrossRef]
9. Sciorio, R.; Cantatore, C.; D'Amato, G.; Smith, G.D. Cryopreservation, cryoprotectants, and potential risk of epigenetic alteration. *J. Assist. Reprod. Genet.* **2024**, *41*, 2953–2967. [CrossRef]
10. Zhang, Y.; Liu, D.; Wang, Q.; Ruan, Q.; Hua, S.; Zhang, W.; Yang, S.; Meng, Z. Addition of cryoprotectant DMSO reduces damage to spermatozoa of yellow catfish (*Pelteobagrus fulvidraco*) during cryopreservation: Ultrastructural damage, oxidative damage and DNA damage. *Animals* **2024**, *14*, 2652. [CrossRef]
11. Kittiphattanabawon, P.; Sriket, C.; Nalinanon, S.; Visessanguan, W.; Benjakul, S. Gelatin hydrolyzed by papaya latex enzymes as an alternative cryoprotectant for frozen raw Pacific white shrimp (*Penaeus vannamei*). *Food Biosci.* **2024**, *60*, 104199. [CrossRef]
12. Anjos, C.; Duarte, D.; Fatsini, E.; Matias, D.; Cabrita, E. Comparative transcriptome analysis reveals molecular damage associated with cryopreservation in *Crassostrea angulata* D-larvae rather than to cryoprotectant exposure. *BMC Genom.* **2024**, *25*, 591. [CrossRef] [PubMed]

13. Xu, Z.; Cao, S.; Zhu, Z.; Hu, B.; Chen, H.; Tu, M.; Tan, Z.; Du, M.; Li, T. Characterization and the mechanism underlying the cryoprotective activity of a peptide from large yellow croaker (*Pseudosciaena crocea*). *Food Chem.* **2024**, *435*, 137512. [CrossRef] [PubMed]
14. Acosta-Salmon, H.; Jerry, D.R.; Southgate, P.C. Effects of cryoprotectant agents and freezing protocol on motility of black-lip pearl oyster (*Pinctada margaritifera* L.) spermatozoa. *Cryobiology* **2007**, *54*, 13–18. [CrossRef]
15. Ishiguro, H.; Rubinsky, B. Influence of fish antifreeze proteins on the freezing of cell suspensions with cryoprotectant penetrating cells. *Int. J. Heat Mass Transf.* **1998**, *41*, 1907–1915. [CrossRef]
16. Upton, R.; Clulow, S.; Colyvas, K.; Mahony, M.; Clulow, J. Paradigm shift in frog sperm cryopreservation: Reduced role for non-penetrating cryoprotectants. *Reproduction* **2023**, *165*, 583–592. [CrossRef]
17. Zhang, M.; Ma, X.; Han, Y.; Wang, Z.; Jia, Z.; Chen, D.; Qiao, Z.; Gao, X.; Zhao, C.; Shen, Y. Optimal conditions for cryopreservation by vitrification of largemouth bass (*Micropterus salmoides*) embryos. *Anim. Reprod. Sci.* **2024**, *270*, 107613. [CrossRef]
18. Gupta, S.; Fletcher, G.C.; Taylor, R.; Hedderley, D.; Cruz, C.D. Effect of air blast freezing and frozen storage on inactivation of *Vibrio vulnificus* in Pacific oysters (*Crassostrea gigas*). *Food Control* **2024**, *163*, 110467. [CrossRef]
19. Yan, L.; Li, Y.; Wang, Z.; Su, J.; Yu, R.; Yan, X.; Ma, P.; Cui, Y. Stress response to low temperature: Transcriptomic characterization in *Crassostrea sikamea* × *Crassostrea angulata* hybrids. *Aquac. Res.* **2018**, *49*, 3374–3385. [CrossRef]
20. Jiang, P.; Qin, X.; Fan, X.; Zhang, C.; Chen, D. The impact of anhydrous storage and transportation at ice temperatures on the post-capture viability and quality of *Patinopecten yessoensis*. *J. Agric. Food Res.* **2024**, *18*, 101277. [CrossRef]
21. Lauren, L.; Dean, R.J.; Paul, C.S. Cryopreservation of black-lip pearl oyster (*Pinctada margaritifera* L.) spermatozoa: Effects of cryoprotectants on spermatozoa motility. *J. Shellfish Res.* **2005**, *24*, 1187–1190. [CrossRef]
22. Claudet, P.V.; Narasimman, S.; Natesan, M. Effect of cryoprotectants and cooling rates on fertility potential of sperm in the giant freshwater prawn, *Macrobrachium rosenbergii* (De Man). *Anim. Reprod. Sci.* **2016**, *171*, 49–57. [CrossRef]
23. Seshoka, M.M.; Mphaphathi, M.L.; Nedambale, T.L. Comparison of four different permitting and combination of two best cryoprotectants on freezing Nguni sperm evaluated with the aid of computer aided sperm analysis. *Cryobiology* **2016**, *72*, 232–238. [CrossRef] [PubMed]
24. Shaliutina-Kolešová, A.; Cosson, J.; Lebeda, I.; Gazo, I.; Shaliutina, O.; Dzyuba, B.; Linhart, O. The influence of cryoprotectants on sturgeon (*Acipenser ruthenus*) sperm quality, DNA integrity, antioxidant responses, and resistance to oxidative stress. *Anim. Reprod. Sci.* **2015**, *159*, 66–76. [CrossRef] [PubMed]
25. Abinawanto, A.; Vardini, N.; Kristanto, A.H.; Lestari, R.; Bowolaksono, A. Effect of egg yolk of free-range chicken and methanol as a cryoprotective agent for the sperm preservation of cyprinid fish, *Neolissochilus soroides* (Valenciennes, 1842). *Heliyon* **2021**, *7*, e08158. [CrossRef]
26. Lee, Y.H.; Park, J.Y.; Lee, I.Y.; Zidni, I.; Lim, H.K. Effects of cryoprotective agents and treatment methods on sperm cryopreservation of stone flounder, *Kareius bicoloratus*. *Aquaculture* **2021**, *531*, 735969. [CrossRef]
27. Ahn, J.Y.; Park, J.Y.; Lim, H.K. Effects of different diluents, cryoprotective agents, and freezing rates on sperm cryopreservation in *Epinephelus akaara*. *Cryobiology* **2018**, *83*, 60–64. [CrossRef] [PubMed]
28. Li, P.; Li, Z.H.; Dzyuba, B.; Hulak, M.; Rodina, M.; Linhart, O. Evaluating the impacts of osmotic and oxidative stress on common carp (*Cyprinus carpio* L.) sperm caused by cryopreservation techniques. *Biol. Reprod.* **2010**, *83*, 852–858. [CrossRef]
29. Vacquier, V.D.; Hamdoun, A. Cold storage and cryopreservation methods for spermatozoa of the sea urchins *Lytechinus pictus* and *Strongylocentrotus purpuratus*. *Dev. Dyn.* **2024**, *253*, 781–790. [CrossRef]
30. Gwo, J.; Arnold, C.R. Cryopreservation of Atlantic croaker spermatozoa: Evaluation of morphological changes. *J. Exp. Zool.* **1992**, *264*, 444–453. [CrossRef]
31. Hassan, M.M.; Li, X.; Liu, Y.; Qin, J.G. Sperm cryopreservation in the spermcasting Australian flat oyster *Ostrea angasi* by a programmable freezing method. *Cryobiology* **2017**, *76*, 119–124. [CrossRef]
32. Yang, H.; Huo, Y.; Yee, J.C.; Rikard, S.; Walton, W.C.; Saillant, E. Sperm repository for a breeding program of the eastern oyster *Crassostrea virginica*: Sample collection, processing, cryopreservation, and data management plan. *Animals* **2021**, *11*, 2836. [CrossRef]
33. Riesco, M.F.; Félix, F.; Matias, D.; Joaquim, S.; Suquet, M.; Cabrita, E. Comparative study on cellular and molecular responses in oyster sperm revealed different susceptibilities to cryopreservation. *Aquaculture* **2019**, *498*, 223–229. [CrossRef]
34. Zheng, X.; Gu, Z.; Huang, Z.; Ding, H.; Vasquez, H.E.; Liu, Y.; Shi, Y.; Wang, A. The effects of cryoprotectants on sperm motility of the Chinese pearl oyster, *Pinctada fucata martensii*. *Cryobiology* **2018**, *82*, 64–69. [CrossRef]
35. Kawamoto, T.; Narita, T.; Isowa, K.; Aoki, H.; Hayashi, M.; Ohta, H.; Komaru, A. Effects of cryopreservation methods on post-thaw motility of spermatozoa from the Japanese pearl oyster, *Pinctada fucata martensii*. *Cryobiology* **2007**, *54*, 19–26. [CrossRef]
36. Arita, K.; Takamatsu, S.; Isowa, K.; Aoki, H.; Ohta, H. Development of a novel non-programmable cryopreservation method capable of accurate cooling rate manipulation. *Aquaculture* **2018**, *484*, 145–151. [CrossRef]
37. Suquet, M.; Gourtay, C.; Donval, A.; Le Goïc, N.; Quere, C.; Malo, F.; Le Grand, J.; Ratiskol, D.; Mingant, C.; Fauvel, C. The quality of great scallop (*Pecten maximus*) sperm after thawing. *Gen. Comp. Endocrinol.* **2016**, *229*, 127–131. [CrossRef]

38. Narita, T.; Kawamoto, T.; Isowa, K.; Aoki, H.; Hayashi, M.; Ohta, H.; Komaru, A. Effects of cryopreservation on sperm structure in Japanese pearl oyster *Pinctada fucata martensii*. *Fish. Sci.* **2008**, *74*, 1069–1074. [CrossRef]
39. Rusk, A.B.; Alfaro, A.C.; Young, T.; Watts, E.; Adams, S.L. Development stage of cryopreserved mussel (*Perna canaliculus*) larvae influences post-thaw impact on shell formation, organogenesis, neurogenesis, feeding ability and survival. *Cryobiology* **2020**, *93*, 121–132. [CrossRef] [PubMed]
40. Zheng, X.; Fu, Z.; Lin, S.; Yang, R.; Wang, A.; Gu, Z.; Ma, Z. Which is the major trigger in aquatic environment for pearl oyster *Pinctada fucata martensii* sperm from gonad: Ammonia ion or pH? *Aquaculture* **2020**, *520*, 734673. [CrossRef]
41. Zeng, Q.; Chen, Y.; Wang, M.; Li, Y.; Dai, T.; Qin, W.; Zhu, Y.; Zhang, C.; Zhou, Y.; Qin, Q.; et al. Study on sperm cryopreservation of hybrid fish derived from *Carassius cuvieri* (♀) × *Carassius auratus* red var. (♂). *Reprod. Breed.* **2024**, *4*, 234–242. [CrossRef]
42. Superio, J.; Resseguier, J.; Nobrega, R.H.; Grebstad, C.M.; Fakriadis, I.; Foss, A.; Hagen, Ø.; Zhang, M.; Del Pilar García-Hernández, M.; Galindo-Villegas, J. Unravelling spermatogenesis in spotted wolffish: Insights from the ultrastructure of juvenile male testes to the cryopreservation of broodstock sperm. *Aquaculture* **2024**, *592*, 741214. [CrossRef]
43. Abdissa, B.; Getahun, A.; Dejen, E. Cryopreservation of sperm of *Labeobarbus brevicephalus* (Pisces: Cyprinidae) from Lake Tana (Ethiopia). *Aquaculture* **2022**, *548*, 737697. [CrossRef]
44. Dhanasekar, K.; Selvakumar, N.; Munuswamy, N. Cryopreservation of sperm in cobia, *Rachycentron canadum* (Linnaeus, 1766). *Aquaculture* **2022**, *557*, 738313. [CrossRef]
45. Handayani, L.; Maulida, S.; Rahayu, S.; Razi, N.M.; Kocabas, M.; Kocabas, F.K.; Wilkes, M.; Siti-Azizah, M.N.; Eriani, K.; Fadli, N.; et al. Effect of cryoprotectant and concentration on the sperm quality of walking catfish, *Clarias batrachus*, post-cryopreservation. *CryoLetters* **2024**, *45*, 320–328. [CrossRef] [PubMed]
46. Hai, E.; Li, B.; Zhang, J.; Zhang, J. Sperm freezing damage: The role of regulated cell death. *Cell Death Discov.* **2024**, *10*, 239. [CrossRef]
47. Ding, X.; Tian, Y.; Qiu, Y.; Duan, P.; Wang, X.; Li, Z.; Li, L.; Liu, Y.; Wang, L. Effects of long-term cryopreservation on the transcriptomes of giant grouper sperm. *Genes* **2024**, *15*, 523. [CrossRef] [PubMed]
48. Pan, J.; Zhu, K.; Liu, J.; Guo, H.; Liu, B.; Zhang, N.; Xian, L.; Sun, J.; Zhang, D. Cryopreservation of goldlined seabream *Rhabdosargus sarba* (Forsskål, 1775) sperm: CASA observation and enzyme activity evaluation. *Aquaculture* **2024**, *582*, 740494. [CrossRef]
49. Balamurugan, R.; Karthik, S.; Arul, V. Effect of cryopreservation on motility, DNA integrity and gene expression in grey mullet, *Mugil cephalus* sperm. *Cryobiology* **2024**, *114*, 104848. [CrossRef] [PubMed]
50. Yang, H.; Hu, E.; Tiersch, T.; Carmichael, C.; Matthews, J.; Varga, Z.M. Temporal and concentration effects of methanol on cryopreservation of zebrafish (*Danio rerio*) sperm. *Zebrafish* **2020**, *17*, 233–242. [CrossRef]
51. Zilli, L.; Bianchi, A.; Sabbagh, M.; Pecoraro, L.; Schiavone, R.; Vilella, S. Development of sea bream (*Sparus aurata*) semen vitrification protocols. *Theriogenology* **2018**, *110*, 103–109. [CrossRef]
52. Kása, E.; Lujić, J.; Marinović, Z.; Kollár, T.; Bernáth, G.; Bokor, Z.; Urbányi, B.; Lefler, K.K.; Jesenšek, D.; Horváth, Á. Development of sperm vitrification protocols for two endangered salmonid species: The Adriatic grayling, *Thymallus thymallus*, and the marble trout, *Salmo marmoratus*. *Fish Physiol. Biochem.* **2018**, *44*, 1499–1507. [CrossRef]
53. Varela, J.A.; Jardim, R.D.; Streit, D.J.; Cardoso, T.F.; Silva, E.F.; Lucia, T.J.; Figuelredo, M.; Corcini, C.D. Trehalose in extenders for cryopreservation of tambaqui (*Colossoma macropomum*) sperm. *CryoLetters* **2022**, *43*, 264–268. [CrossRef]
54. Pandey, D.; Ryu, Y.; Matsubara, T. Features of sperm motility and circadian rhythm in Japanese anchovy (*Engraulis japonicus*). *Fish. Aquac. J.* **2017**, *8*, 203. [CrossRef]
55. Cosson, J.; Groison, A.L.; Suquet, M.; Fauvel, C.; Dreanno, C.; Billard, R. Studying sperm motility in marine fish: An overview on the state of the art. *J. Appl. Ichthyol.* **2008**, *24*, 460–486. [CrossRef]
56. França, T.S.; González-López, W.A.; Sanchez, M.P.; Ferrão, L.; Fernández-García, F.; Borges, L.P.; Belenguer, A.; Holorea, P.G.; Calduch-Giner, J.C.; Felip, A.; et al. Successful cryopreservation in biodegradable containers of sperm from aquaculture Mediterranean fishes. *Theriogenology* **2024**, *216*, 53–61. [CrossRef]
57. Salinas-Flores, L.; Paniagua-Chavez, C.G.; Jenkins, J.A.; Tiersch, T.R. Cryopreservation of sperm of red abalone (*Haliotis rufescens*). *J. Shellfish Res.* **2005**, *24*, 415–420. [CrossRef]
58. Ferrão, L.; Morini, M.; González-Lopéz, W.A.; Galledo, V.; Felip, A.; Pérez, L.; Asturiano, J.F. Effects of cold seawater pre-treatments on induction of early sexual maturation and sperm production in European eel (*Anguilla anguilla*). *Fish Physiol. Biochem.* **2024**, *50*, 2489–2503. [CrossRef] [PubMed]
59. Shaliutina-Loginova, A.; Loginov, D.S. Oxidative stress and DNA fragmentation in frozen/thawed common carp (*Cyprinus carpio*) sperm with and without supplemental proteins. *Anim. Reprod. Sci.* **2023**, *251*, 107213. [CrossRef]
60. Stanger, S.J.; Law, E.A.; Jamsai, D.; O'Bryan, M.K.; Nixon, B.; McLaughlin, E.A.; Aitken, R.J.; Roman, S.D. A novel germ cell protein, SPIF (sperm PKA interacting factor), is essential for the formation of a PKA/TCP11 complex that undergoes conformational and phosphorylation changes upon capacitation. *FASEB J.* **2016**, *30*, 2777–2791. [CrossRef]
61. Xin, M.; Siddique, M.A.M.; Dzyuba, B.; Cuevas-Uribe, R.; Shaliutina-Kolešová, A.; Linhart, O. Progress and challenges of fish sperm vitrification: A mini review. *Theriogenology* **2017**, *98*, 16–22. [CrossRef] [PubMed]

62. Hu, Y.; Wang, X.; Liu, X.; Zhou, Y.; Zheng, J.; Zheng, X. Sperm cryopreservation of *Amphioctopus fangsiao* (Mollusca: Cephalopoda). *Aquaculture* **2025**, *598*, 742026. [CrossRef]
63. Kiriyakit, A.; Gallardo, W.G.; Bart, A.N. Successful hybridization of groupers (*Epinephelus coioides* × *Epinephelus lanceolatus*) using cryopreserved sperm. *Aquaculture* **2011**, *320*, 106–112. [CrossRef]

Journal of
*Marine Science
and Engineering*

MDPI

Article

Comparative Analysis of Amino Acid Composition, Fatty Acid Profiles, and Genetic Diversity Among Three Populations of *Penaeus semisulcatus*

Yundong Li [1,2,3,†], Juan Chen [1,†], Ziyi Jiang [4], Song Jiang [1], Qibin Yang [1,2], Lishi Yang [1], Jianhua Huang [1,3], Jianzhi Shi [1], Yangyang Ding [1], Tianmi Liu [5,*] and Falin Zhou [1,2,3,*]

1 South China Sea Fisheries Research Institute, Chinese Academy of Fishery Sciences, Key Laboratory of South China Sea Fishery Resources Exploitation and Utilization, Ministry of Agriculture and Rural Affairs, Guangzhou 510300, China; liyd2019@163.com (Y.L.); chenj29725@163.com (J.C.); tojiangsong@163.com (S.J.); yangqibin1208@163.com (Q.Y.); yangls2016@163.com (L.Y.); huangjianhua@scsfri.ac.cn (J.H.); shijianzhi1989@163.com (J.S.); dingyangyang93@163.com (Y.D.)
2 Key Laboratory of Efficient Utilization and Processing of Marine Fishery Resources of Hainan Province, Sanya Tropical Fisheries Research Institute, Sanya 572018, China
3 Shenzhen Base of South China Sea Fisheries Research Institute, Chinese Academy of Fishery Sciences, Shenzhen 518108, China
4 Keck School of Medicine, University of Southern California, 1975 Zonal Ave, Los Angeles, CA 90033, USA; ziyijian@usc.edu
5 Fisheries Technology Extension Center of Hainan Province, Haikou 570311, China
* Correspondence: ltm600@163.com (T.L.); zhoufalin@scsfri.ac.cn (F.Z.)
† These authors contributed equally to this work.

Abstract: *Penaeus semisulcatus*, a commercially valuable aquaculture species, is widely distributed along the southeastern coast of China and throughout the South China Sea. Despite its economic importance, the comprehensive understanding of its germplasm characteristics remains limited. This investigation evaluated nutrient profiles, amino acid composition, lipid constituents, antioxidant capacity, and genomic variability across three farmed aquaculture populations maintained under standardized environmental parameters to discern divergence in nutritive attributes and hereditary characteristics. The results revealed significant interpopulation variation in ash content ($p < 0.05$), while other proximate components showed comparable levels. Among 17 detected amino acids, glutamic acid was the most abundant (2.07–2.30 g/100 g), while cysteine had the lowest concentration (0.10–0.11 g/100 g). Notably, the PsHK population had a relatively higher amino acid content, indicating superior nutritional value. Fatty acid analysis detected 13 fatty acids, with C18:0 (88.27–108.00 mg/100 g) being the most abundant, and C20:2 (3.40–4.70 mg/100 g) the least. The PsHK population exhibited significantly higher levels of all fatty acids compared to the other two populations ($p < 0.05$). Antioxidant enzyme activity assays revealed no significant differences in antioxidant capacity across the three populations, indicating a similar overall antioxidant status. Genetic diversity analysis indicated that the PsHK population had the highest genetic diversity, with a relatively pronounced genetic differentiation between the PsHK and PsRP populations. In conclusion, the comparative analysis of these three cultured populations highlights that the PsHK population excels in both nutritional composition and genetic diversity. With its superior nutritional profile and rich genetic background, *P. semisulcatus* shows great potential for aquaculture development. These findings provide valuable insights for future germplasm improvement and aquaculture optimization efforts, offering a scientific basis for refining breeding strategies and enhancing the nutritional evaluation of *P. semisulcatus*. Additionally, the comparative analysis of genetic diversity and biochemical composition contributes to a better understanding of population variation, which is essential for the sustainable management and utilization of this species in aquaculture.

Keywords: *Penaeus semisulcatus*; amino acids; fatty acids; antioxidant enzyme activity; genetic diversity

1. Introduction

The green tiger shrimp (*Penaeus semisulcatus*) is a widely distributed species that can be found across an extensive range, stretching from the southwestern Indian Ocean to the Indo-Malayan archipelago. Its habitat spans diverse regions such as the Erythraean Sea, Arabian Gulf, and the western littoral zones of Madagascar, reflecting its adaptability to various coastal environments [1]. This species is considered one of the most significant commercial shrimp in the Indo-Pacific region, with notable economic importance along the eastern Mediterranean coast as well [2]. In China, *P. semisulcatus* is primarily found along the southeastern coast, particularly in the provinces of Fujian, Taiwan, Guangdong, Guangxi, and Hainan Island. Within Hainan, it is most commonly distributed in the eastern and southern waters, with fewer occurrences along the western and northern coasts. This shrimp is prized for its fast growth, large size, and substantial economic value.

P. semisulcatus, as a highly promising aquaculture species, has limited current research focusing mainly on reproductive biology [3,4], population dynamics [5], and other related areas. In contrast, research on species such as *Penaeus monodon* [6], *Litopenaeus vannamei* [7], and *Fenneropenaeus chinensis* [8] is extensive and in-depth, with several studies on their nutritional aspects. However, studies on *P. semisulcatus* are relatively limited. Researchers evaluating the nutritional value of male and female shrimp from the Mediterranean (Alexandria) and Indian markets identified 16 amino acids and 15 fatty acids. Their findings showed that the edible muscle fat content of both male and female shrimp from Alexandria was higher than that of the same species from the Indian market. Additionally, the carbohydrate content in the muscle of female shrimp was lower compared to that of males [9]. In terms of genetic diversity, the research on *P. semisulcatus* is relatively broader. A genetic diversity analysis of populations from the Indo-Pacific region revealed that the species can be divided into two major lineages, with a distinct genetic boundary near Bagan Pasir (BGP) in the Strait of Malacca [1]. Additionally, studies on the genetic structure of populations in the Persian Gulf showed significant genetic differentiation among groups, with some loci deviating from Hardy–Weinberg equilibrium and others exhibiting null alleles, suggesting the presence of genetic bottlenecks or selective pressure within the populations [10]. Furthermore, using the RAPD method to analyze the genetic diversity and differentiation of two wild populations of *P. semisulcatus* from Xiamen and Shantou coasts in China, the results indicated high genetic diversity and low genetic differentiation, suggesting that these populations may belong to the same group and hold potential for further development [11].

While previous studies have offered initial insights into the germplasm resources and genetic diversity of *P. semisulcatus*, research in this area remains scarce, especially in China, where region-specific studies are still lacking. To address this gap, the present study selected three representative aquaculture regions in China-Haikou, Raoping, and Sanya—for systematic sample collection. Muscle tissue samples of *P. semisulcatus* were gathered from these areas to assess their nutritional value, and a comprehensive genetic resource database was established. To assess population-level genomic variability, we conducted molecular profiling of *P. semisulcatus* across sampled geographic zones. By integrating these analyses, we provide empirical evidence to support science-driven conservation planning, ensuring both ecological sustainability and the continued productivity of local fisheries. The methodological framework developed in this study offers valuable, transferable solutions for the preservation and sustainable management of marine organisms, providing a foundation for similar conservation initiatives in marine aquaculture.

2. Materials and Methods

2.1. Sample Collection

In specific aquaculture regions for *P. semisulcatus* in China, we randomly selected three representative farming areas: Haikou, Raoping, and Sanya (Table 1, Figure 1). Accurately calibrated instruments, including pH meters, dissolved oxygen sensors, and salinity meters, along with Merck chemical test kits using colorimetric analysis, were employed to measure various aquatic parameters such as salinity, ammonium-N, nitrite levels, pH, dissolved oxygen, and nitrate concentrations. The statistical analysis using ANOVA and *t*-tests showed significant consistency in these metrics among different breeding groups, further indicating that the aquaculture environments across the surveyed areas are comparable. Notably, recorded water temperatures varied between 27.4 °C and 28.5 °C at measurement points, maintaining stable alkaline conditions with pH levels approximating 8.8. Dissolved oxygen (DO) levels were consistently maintained above 5 mg/L throughout the study. Salinity was carefully monitored using a salinity meter to ensure it remained at 30‰. Additionally, we regularly measured ammonia, nitrite, and nitrate levels to ensure water quality, maintaining ammonia levels below 0.1 mg/L, nitrite below 0.2 mg/L, and nitrate below 50 mg/L at all times. Ninety specimens (approximately 50% males and 50% females) were randomly collected from three aquaculture sites, with thirty specimens selected from each site. These 30 specimens per site were first used to measure body length, weight and genetic diversity. Following this, for the subsequent analyses of amino acid profiling, fatty acid composition and enzyme activity, three individuals were randomly selected from each group for further testing. Muscle tissues (1 g aliquots) were aseptically excised from each organism, transferred to sterile centrifuge tubes, and cryopreserved at −80 °C for subsequent molecular investigations.

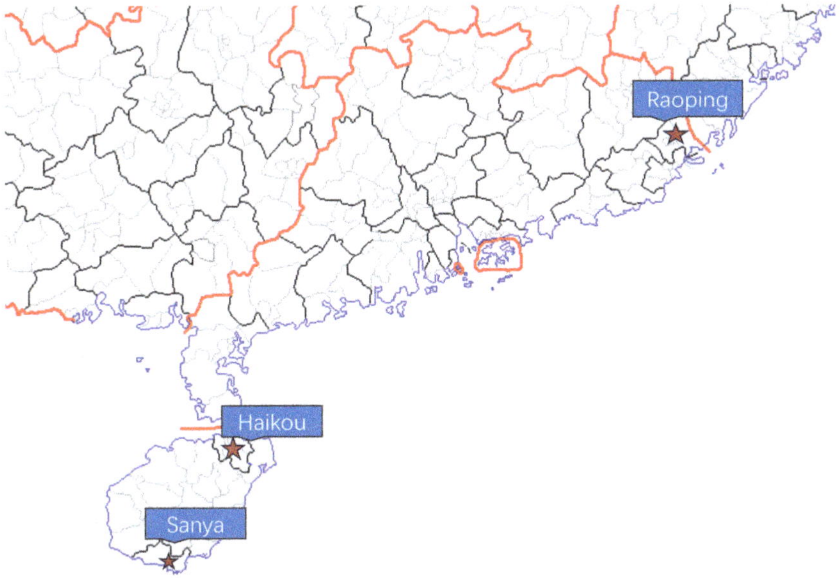

Figure 1. Map of the survey point. (On the map, red lines mark provincial boundaries, blue represents the coastline, and the circled areas highlight the Special Administrative Regions of Hong Kong and Macao in China.)

Table 1. Basic information of the sampling sites.

Population	Body Length (mm)	Weight (g)	Location
Haikou, Hainan Province (PsHK)	155.4 ± 13.6	26.95 ± 5.46	20°03′ N, 110°37′ E
Raoping, Guangdong Province (PsRP)	140.8 ± 12.3	24.37 ± 6.52	23°66′ N, 117°00′ E
Sanya, Hainan Province (PsSY)	135.6 ± 16.7	22.39 ± 6.43	18°25′ N, 109°51′ E

Morphometric parameters (body length and mass) were quantified as arithmetic mean ± SD ($n = 30$).

2.2. Comprehensive Nutritional Analysis

Moisture quantification was performed in compliance with GB 5009.3-2016 [12] standard protocols, employing precision oven-drying methodology. Pre-weighed samples in calibrated crucibles underwent thermal dehydration at 105 °C for 240 min, followed by vacuum desiccator equilibration and gravimetric measurement through sequential drying cycles until mass stabilization. Ash determination followed GB 5009.4-2016 [13] muffle furnace (Thermo Scientific Thermolyne, Waltham, MA, USA) protocols, while protein content (GB 5009.5-2016 [14]) and lipid fractions (GB/T 5009.6-2016 [15]) were analyzed via micro-Kjeldahl (Kjeltec 8400, Hillerød, Denmark) distillation and automated Soxhlet (LabTech LT-01, Beijing, China) extraction systems, respectively. Carbohydrate composition was established per GB/T 15672-2009 [16] specifications using sulfuric acid–phenol chromogenic reactions with spectrophotometric detection.

2.3. Nutritional and Antioxidant Analysis

Amino acid profiling was conducted per GB 5009.124-2016 [17] specifications through reverse-phase HPLC separation with UV detection (440 nm for proline (Pro) and 570 nm for other amino acids). Lipid characterization was conducted following the GB 5009.168-2016 [18] guidelines, using certified reference materials for the quantitative analysis of fatty acid methyl esters (FAMEs). The preparation of FAMEs involved an acid-catalyzed methylation process, where fatty acids were esterified with methanol in the presence of sulfuric acid as the catalyst, ensuring complete methylation for accurate analysis. Antioxidant biomarkers were evaluated using standardized commercial kits (Nanjing Jiancheng Bioengineering Institute, Nanjing, China): Superoxide Dismutase (SOD) activity via WST-8 redox cycling (A001-1-2), Total Antioxidant Capacity (T-AOC) through ferric reducing power assay (A015-1), and Catalase (CAT) enzymatic conversion using ammonium molybdate chromogenic system (A007-1-1).

2.4. Genetic Diversity

DNA samples that met the quality criteria were sequenced using the MGISEQ series sequencer (MGI Tech Co., Ltd., Shenzhen, China). In this study, a total of 90 individuals were sequenced, generating a total of 2562.36 Gb of data. On average, each sample produced 28.47 Gb of sequencing data, which corresponds to an average sequencing depth of 14.23X. After read alignment, we observed that the average alignment rate to the reference genome was 64.88%, with an average sequencing depth of 50.67X across the aligned reads. The process involved random fragmentation, fragment selection, end repair, adapter ligation, PCR preparation of DNA nanoballs (DNBs), and sequencing on a patterned sequencing chip. The sequencing quality was assessed using Phred scores, and Raw sequencing reads underwent quality control processing via fastp (v0.23.2) with adapter trimming and quality filtering (Q20 > 90%, read length ≥ 150 bp), generating high-fidelity data suitable for downstream genomic analyses. Next, clean data were aligned to the reference genome using BWA (Wellcome Trust Sanger Institute, Cambridge, UK), with re-alignment performed around InDel edge regions. SNP detection was performed using GATK's HaplotypeCaller (Broad Institute, GATK 4.x, https://gatk.broadinstitute.org/), generating mutation files in VCF format with variant annotations. For multi-sample analysis, gVCF files were first generated for each individual sample and subsequently processed using GATK's GenotypeGVCFs (Broad Institute, GATK 4.x, https://gatk.broadinstitute.org/) tool for joint genotyping, enabling the identification of population-level genetic variations.

2.5. Data Analysis

Quantitative data processing was executed in SPSS (v16.0) through hierarchical analytical protocols. Primary statistical evaluation employed one-way ANOVA to detect intergroup variations, followed by Duncan's post-hoc testing when omnibus significance thresholds ($\alpha = 0.05$) were attained, establishing pairwise differential significance. All values are expressed as $x \pm s$. In our analysis, various indices were used to measure genetic variation, including SNP density = SNP count/genome size (Kb), Ho = het/(het + hom), PIC = $1 - \Sigma pi^2 - \Sigma\Sigma 2pi^2pj^2$, $\pi = n/(n-1)\ \Sigma pipj\pi ij$, F = (Ho − He)/($n$ − He).

3. Results

3.1. Analysis of Overall Nutritional Composition

A routine nutritional analysis in the three *P. semisulcatus* aquaculture populations (Table 2) revealed the following ranges: moisture (74.07–74.57 g/100 g), ash (1.77–2.07 g/100 g), crude fat (0.80–0.83 g/100 g), crude protein (21.77–22.37 g/100 g), and total sugar (29.92–30.92%). Overall, the nutritional composition showed minimal variation among the populations, with no significant differences in moisture, crude protein, crude fat, or total sugar content ($p > 0.05$). However, ash content varied significantly ($p < 0.05$), with the PsRP population having the highest value (2.07 g/100 g), significantly differing from PsHK and PsSY ($p < 0.05$). In contrast, no significant difference was found between PsHK (1.77 g/100 g) and PsSY (1.80 g/100 g) ($p > 0.05$).

Table 2. Nutrient profile in the myotomal tissues across three populations of *P. semisulcatus*.

General Nutrition Content	PsHK	PsRP	PsSY
Moisture (g/100 g)	74.07 ± 0.71^a	74.50 ± 0.70^a	74.57 ± 0.55^a
Ash (g/100 g)	1.77 ± 0.06^a	2.07 ± 0.06^b	1.80 ± 0.10^a
Crude Fat (g/100 g)	0.83 ± 0.06^a	0.80 ± 0.10^a	0.83 ± 0.06^a
Crude Protein (g/100 g)	22.37 ± 0.9^a	21.77 ± 0.85^a	22.13 ± 0.47^a
Total Sugar (%)	30.92 ± 2.47^a	30.70 ± 3.69^a	29.92 ± 1.72^a

Mean ± standard deviation ($n = 3$). Superscript letters (a, b) in the same row indicate statistically significant differences between groups, while identical letters denote no significant difference ($p > 0.05$).

3.2. Analysis of Amino Acids

Chromatographic quantification of muscle amino acids in triplicate *P. semisulcatus* stocks (Table 3) resolved 17 proteogenic components, comprising seven essential (EAA), two semi-essential (SEAA), and eight non-essential (NEAA) variants. Multivariate analysis revealed statistical divergences ($p < 0.05$) in 13 amino acid concentrations across cohorts, whereas Thr, Ser, Cys, and Arg demonstrated conserved concentrations ($p > 0.05$) irrespective of population origin.

Table 3. Amino acid composition and levels in the myotomal tissues of the three *P. semisulcatus* populations.

Content (g/100 g)	PsHK	PsRP	PsSY
Thr *	0.58 ± 0.01 ^{a}	0.58 ± 0.02 ^{a}	0.54 ± 0.02 ^{a}
Val *	0.73 ± 0.01 ^{a}	0.72 ± 0.03 ^{a}	0.65 ± 0.03 ^{b}
Met *	0.35 ± 0.01 ^{a}	0.33 ± 0.02 $^{a b}$	0.31 ± 0.01 ^{b}
Ile *	0.67 ± 0.01 ^{a}	0.66 ± 0.03 $^{a b}$	0.60 ± 0.03 ^{b}
Leu *	1.22 ± 0.01 ^{a}	1.20 ± 0.04 $^{a b}$	1.13 ± 0.05 ^{b}
Phe *	0.62 ± 0.01 ^{a}	0.59 ± 0.02 ^{a}	0.55 ± 0.02 ^{b}
Lys *	1.32 ± 0.02 ^{a}	1.27 ± 0.05 $^{a b}$	1.21 ± 0.04 ^{b}
EAA	5.49 ± 0.05 ^{a}	5.36 ± 0.19 $^{a b}$	4.99 ± 0.18 ^{b}
Asp @	1.54 ± 0.02 ^{a}	1.51 ± 0.06 ^{a}	1.39 ± 0.05 ^{b}
Glu @	2.30 ± 0.03 ^{a}	2.24 ± 0.08 ^{a}	2.07 ± 0.08 ^{b}
Gly @	1.13 ± 0.01 ^{a}	1.42 ± 0.05 ^{b}	1.47 ± 0.06 ^{b}
Ala @	0.91 ± 0.01 ^{a}	0.90 ± 0.03 ^{a}	0.83 ± 0.03 ^{b}
DAA	5.89 ± 0.06 ^{a}	6.08 ± 0.22 ^{a}	5.77 ± 0.21 ^{a}
His &	0.30 ± 0.01 ^{a}	0.29 ± 0.02 ^{a}	0.26 ± 0.01 ^{b}
Arg &	1.65 ± 0.03 ^{a}	1.68 ± 0.05 ^{a}	1.63 ± 0.05 ^{a}
SEAA	1.95 ± 0.03 ^{a}	1.98 ± 0.07 ^{a}	1.89 ± 0.06 ^{a}
Ser	0.49 ± 0.01 ^{a}	0.48 ± 0.01 ^{a}	0.46 ± 0.03 ^{a}
Cys	0.11 ± 0.01 ^{a}	0.11 ± 0.02 ^{a}	0.10 ± 0.01 ^{a}
Tyr	0.30 ± 0.01 ^{a}	0.24 ± 0.01 ^{b}	0.28 ± 0.01 ^{a}
Pro	0.80 ± 0.01 ^{a}	0.63 ± 0.02 ^{b}	0.51 ± 0.02 ^{c}
NEAA	9.51 ± 0.15 ^{a}	9.54 ± 0.31 ^{a}	9.01 ± 0.32 ^{a}
TAA	15.00 ± 0.20 ^{a}	14.90 ± 0.50 ^{a}	14.00 ± 0.50 ^{a}

Mean ± standard deviation (n = 3). Superscript letters (a, b, c) within the same row denote statistically significant differences between groups. Values sharing the same letter are not significantly different ($p > 0.05$). @ Delicious amino acids. * Essential amino acids. & Semi-essential amino acids. EAA: essential amino acids. DAA: delicious amino acids. SEAA: total semi-essential amino acids. NEAA: total non-essential amino acids. TAA: total amino acids. Thr—Threonine, Val—Valine, Met—Methionine, Ile—Isoleucine, Leu—Leucine, Phe—Phenylalanine, Lys—Lysine, Asp—Aspartic acid, Glu—Glutamic acid, Gly—Glycine, Ala—Alanine, His—Histidine, Arg—Arginine, Ser—Serine, Cys—Cysteine, Tyr—Tyrosine, Pro—Proline.

Among the 17 amino acids, Glu had the highest concentration (2.07–2.30 g/100 g). The PsSY population had the lowest Glu content (2.07 g/100 g), which was significantly lower than that of PsHK (2.30 g/100 g) and PsRP (2.24 g/100 g) ($p < 0.05$). In contrast, Cys had the lowest concentration (0.10–0.11 g/100 g), with no significant differences among the populations ($p > 0.05$). Overall, amino acid levels were highest in PsHK, followed by PsRP, with PsSY having the lowest content.

Comparison of essential amino acids (EAA) among the three populations revealed significant differences ($p < 0.05$). The total EAA content ranged from 4.99 to 5.49 g/100 g, with PsHK exhibiting the highest levels. Additionally, delicious amino acids (DAA) ranged from 5.77 to 6.08 g/100 g, semi-essential amino acids (SEAA) from 1.89 to 11.98 g/100 g, and non-essential amino acids (NEAA) from 9.01 to 9.54 g/100 g. The total amino acid (TAA) content varied between 14.00 and 15.00 g/100 g. Although no significant differences were observed in total amino acid content among the populations ($p > 0.05$), significant variations were noted in the EAA/TAA and SEAA ratios between groups ($p < 0.05$) (Figure 2).

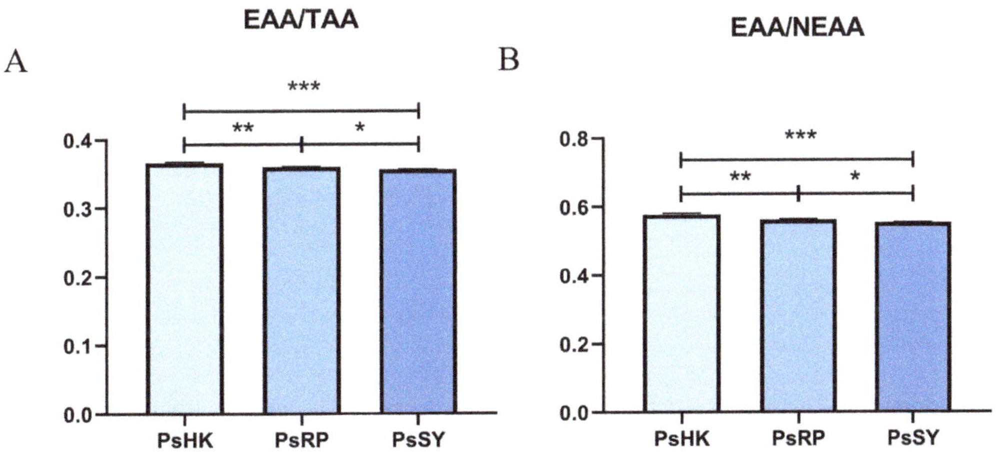

Figure 2. Amino acid composition and levels in the myotomal tissues of the three *P. semisulcatus* populations (n = 3). (**A**): EAA/TAA: essential amino acids/total amino acids. (**B**): EAA/NEAA: essential amino acids/total non-essential amino acids. Different numbers of symbols between treatments indicate significant differences (* $0.01 < p < 0.05$; ** $p < 0.01$, *** $p < 0.001$).

3.3. Analysis of Fatty Acids

Fatty acid analysis in the three *P. semisulcatus* aquaculture populations revealed 13 identified fatty acids (Table 4). C18:0 had the highest concentration (88.27–108.00 mg/100 g), while C20:2 had the lowest (3.40–4.70 mg/100 g). Except for C22:1n9, the PsHK population exhibited the highest levels of all other fatty acids, showing significant differences from the other two populations ($p < 0.05$). In contrast, PsRP and PsSY displayed varying fatty acid contents, but most differences were not significant ($p > 0.05$), indicating similar fatty acid compositions.

Table 4. Muscle fatty acid composition and content of three groups of *P. semisulcatus*.

Content (mg/100 g)	PsHK	PsRP	PsSY
C16:0	70.57 ± 2.99 ^{a}	57.00 ± 2.17 ^{b}	52.73 ± 2.66 ^{b}
C17:0	15.17 ± 0.67 ^{a}	12.67 ± 0.51 ^{b}	12.07 ± 0.64 ^{b}
C18:0	108.00 ± 4.55 ^{a}	88.27 ± 3.31 ^{b}	94.13 ± 4.74 ^{b}
C20:0	9.63 ± 0.35 ^{a}	7.83 ± 0.31 ^{b}	8.13 ± 0.40 ^{b}
C22:0	19.13 ± 0.81 ^{a}	10.90 ± 0.36 ^{b}	12.13 ± 0.58 ^{b}
Saturated fatty acids	222.50 ± 9.37 ^{a}	176.70 ± 6.65 ^{b}	179.80 ± 9.03 ^{b}
C16:1	18.47 ± 0.76 ^{a}	13.13 ± 0.50 ^{b}	13.60 ± 0.70 ^{b}
C18:1n9c	72.80 ± 3.06 ^{a}	50.60 ± 1.87 ^{b}	45.50 ± 2.54 ^{b}
Monounsaturated fatty acids	91.27 ± 3.82 ^{a}	63.73 ± 2.37 ^{b}	59.10 ± 2.95 ^{b}
C18:2n6c	11.23 ± 0.45 ^{a}	8.67 ± 0.35 ^{b}	6.80 ± 0.26 ^{c}
C20:2	4.70 ± 0.26 ^{a}	4.60 ± 0.26 ^{a}	3.40 ± 0.01 ^{b}
C20:4n6	43.70 ± 1.87 ^{a}	35.80 ± 1.74 ^{b}	41.50 ± 2.08 ^{a}
C20:5n3 (EPA)	52.43 ± 2.05 ^{a}	43.53 ± 1.72 ^{b}	47.10 ± 2.29 ^{b}
C22:6n3 (DHA)	70.73 ± 2.87 ^{a}	61.73 ± 2.24 ^{b}	56.63 ± 3.03 ^{b}
DHA + EPA	123.20 ± 4.92 ^{a}	105.3 ± 3.95 ^{b}	103.70 ± 5.31 ^{b}
n-6ΣPUFA	54.93 ± 2.31 ^{a}	44.47 ± 2.08 ^{b}	48.30 ± 2.34 ^{b}
Polyunsaturated fatty acids	182.80 ± 7.33 ^{a}	154.30 ± 6.37 ^{b}	153.20 ± 9.60 ^{b}
C22:1n9	10.13 ± 0.55 ^{a}	10.90 ± 0.72 ^{a,b}	12.07 ± 0.64 ^{b}
Total fatty acids	506.70 ± 20.92 ^{a}	405.60 ± 15.99 ^{b}	403.50 ± 22.21 ^{b}

Mean fatty acids ± standard deviation (n = 3). Different letters between treatments indicate significant differences ($p < 0.05$). C16:0—Palmitic acid. C16:1—Palmitoleic acid. C17:0—Heptadecanoic acid. C18:0—Stearic acid. C18:1n9c—Oleic acid. C18:2n6c—Linoleic acid. C20:0: Arachidic acid. C20:2—Eicosadienoic acid. C22:0—Behenic acid. C20:4n6—Arachidonic acid. C22:1n9—Erucic acid. C20:5n3 (EPA)—Eicosapentaenoic acid. C22:6n3 (DHA)—Docosahexaenoic acid. n-6ΣPUFA: n-6 series polyunsaturated fatty acids.

When comparing the contents of total fatty acids, saturated fatty acids, monounsaturated fatty acids, and polyunsaturated fatty acids, the PsHK population had the highest content in all these categories, with significant differences from the PsRP and PsSY populations ($p < 0.05$). However, no significant difference was observed between the PsRP and PsSY populations. Specifically, the saturated fatty acid content in the three populations ranged from 176.70–222.50 mg/100 g, the monounsaturated fatty acid content ranged from 59.10–91.27 mg/100 g, and the total fatty acid content ranged from 403.50–506.70 mg/100 g.

3.4. Analysis of Antioxidant Enzymes

In the assessment of the activity of three antioxidant enzymes (T-AOC, SOD, and CAT) across the three *P. semisulcatus* populations, the results showed no significant differences among the populations ($p > 0.05$) (Figure 3). Specifically, the T-AOC activity ranged from 2.67–2.68 μmol/g, SOD activity ranged from 50.07–54.96 μmol/(min·g), and CAT activity ranged from 762.7–807.8 U/g. Among these three enzyme activities, the PsHK population generally exhibited higher enzyme activity, indicating stronger antioxidant capacity. In contrast, the PsSY population showed the lowest levels of T-AOC and SOD activities, suggesting weaker antioxidant capacity. Notably, the PsRP population displayed the lowest CAT activity at 762.7 U/g, which was significantly lower than the other two populations, indicating a relatively reduced ability to scavenge peroxides.

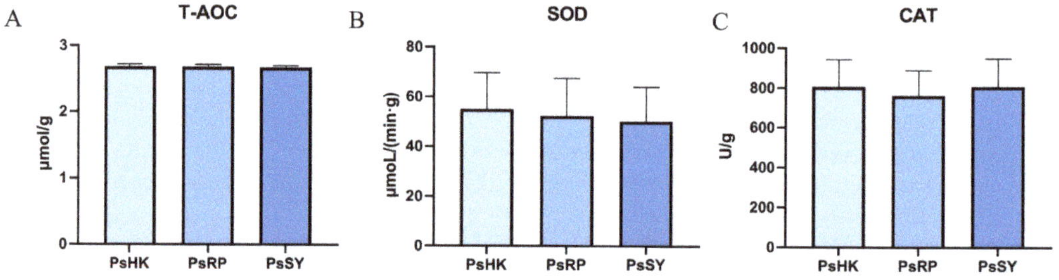

Figure 3. Physiological and biochemical indicators in the muscle of the three *P. semisulcatus* groups. (**A**): The activity of T-AOC. (**B**): The activity of SOD. (**C**): The activity of CAT. Mean ± standard deviation ($n = 3$).

3.5. Genetic Diversity Analysis

The genetic diversity analysis of three cultured *P. semisulcatus* populations revealed that the average SNP density ranged from 1.541–2.109 (Table 5). Nucleotide diversity (π) values were similar among populations, ranging from approximately 3.37×10^{-4} to 4.50×10^{-4}. The inbreeding coefficient (F_{HOM}) ranged from 6.20% to 9.21%, with the PsRP population showing the lowest value, indicating the least risk of inbreeding. The average polymorphic information content (PIC) ranged from 0.094 to 0.129, indicating the genetic diversity of the markers employed. The observed heterozygosity (Ho) had an average value between 0.118 and 0.130, reflecting the frequency of heterozygous individuals within the populations.

Table 5. Statistics of genetic diversity parameters of three populations of *P. semisulcatus*.

Population	SNP Density (SNP/Kb)	Nucleotide (π)	Polymorphism Information Content (PIC)	Observed Heterozygosity (Ho)	Inbreeding Coefficient (F_{HOM})
PsHK	2.109	$4.50 \times 10^{-4} \pm 5.99 \times 10^{-4}$	0.118 ± 0.114	0.120 ± 0.157	$6.42 \times 10^{-2} \pm 5.67 \times 10^{-2}$
PsRP	1.986	$4.49 \times 10^{-4} \pm 5.92 \times 10^{-4}$	0.129 ± 0.114	0.130 ± 0.157	$9.20 \times 10^{-2} \pm 5.16 \times 10^{-2}$
PsSY	1.541	$3.37 \times 10^{-4} \pm 4.45 \times 10^{-4}$	0.094 ± 0.096	0.118 ± 0.095	$6.21 \times 10^{-2} \pm 3.99 \times 10^{-2}$

4. Discussion

4.1. Comparative Analysis of Fundamental Nutritional Components in P. semisulcatus Populations

Water is a fundamental component for all living organisms, and for aquatic species, the regulation of water balance—between intake and loss—is primarily governed by osmotic processes. Aquatic organisms regulate their internal osmotic pressure to ensure equilibrium with the surrounding environment and maintain proper water balance [19]. Additionally, minerals in the ash content are important components that make up the skeleton and hard shells of aquatic organisms [20] and are also essential trace elements for maintaining physiological functions [21]. Protein is the material basis of life and an important component of all living organisms, playing a crucial role in biological activities [22]. Crude protein is an estimate of the nitrogen content in protein, mainly providing energy, forming the body, and regulating physiological functions [23]. Crude fat is the energy storage form in organisms [24] and also a part of cell membranes. In aquatic organisms, crude fat helps maintain body temperature and protects internal organs [25]. Total sugars are one of the main energy sources in organisms, providing rapid energy for aquatic organisms to support their activity and growth [26,27]. Therefore, in *P. semisulcatus*, these routine nutrients are essential components of their bodily functions and primary energy sources.

In the analysis of the routine nutritional components of *P. semisulcatus*, it was found that the crude fat content ranged from 0.80 to 0.83 g/100 g, which is lower than that of *L. vannamei* (0.8–1.1%) [28], *Marsupenaeus japonicus* (2.09 g/100 g) [29], and *Metapenaeus ensis* (0.83–1.03 g/100 g) [30]. Meanwhile, the crude protein content of *P. semisulcatus* ranged from 21.77–22.37 g/100 g, which is higher than that of *Fenneropenaeus merguiensis* (20.83–21.97 g/100 g) [31], *M. japonicus* (14.60 g/100 g), *Procambarus clarkii* (12.33 g/100 g), and *P. monodon* (13.29 g/100 g) [32]. Compared to *P. chinensis* (1.69 g/100 g) and *L. vannamei* (1.52 g/100 g) [33], the ash content of *P. semisulcatus* (1.77–2.07 g/100 g) is relatively higher, with significant differences among the populations ($p < 0.05$). Overall, *P. semisulcatus* is a protein-rich, low-fat food. Among the three populations of *P. semisulcatus*, the PsRP population had relatively low crude fat content but higher ash content, which may give it an advantage in terms of routine nutritional components.

4.2. Comparative Analysis of Amino Acid Profiles in P. semisulcatus Populations

Amino acids in aquatic organisms serve several critical functions. They are fundamental building blocks of proteins and act as precursors to various essential biomolecules, including enzymes, hormones, and antibodies. These molecules are crucial for growth and development, and they play vital roles in osmoregulation, hypoxia tolerance, shell formation, metabolic processes, antioxidant defense, and immune response [34–36]. Amino acids assist aquatic organisms in regulating their internal osmotic pressure, adapting to environments with fluctuating salinity, and enhancing their ability to survive under hypoxic conditions [37,38]. Additionally, they are involved in synthesizing important biomolecules, such as hormones and neurotransmitters, and play a role in antioxidant and immune responses, thus enhancing the overall health and survival ability of aquatic organisms [34]. Furthermore, certain amino acids can serve as an energy source, providing support during stress responses or periods of hunger [39,40]. The balance and metabolism of amino acids have significant effects on the growth, development, and reproduction of aquatic organisms.

In this study, EAA content in *P. semisulcatus* ranged from 4.99 to 5.49 g/100 g, which is lower than that of *P. monodon* (5.44–6.41 g/100 g) [6]. Among these, glutamic acid had the highest content, while cysteine had the lowest, a trend similar to that seen in *M. ensis* [30]. The total amino acid content in *P. semisulcatus* ranged from 14.00–15.00 g/100 g, slightly lower than that in *P. monodon* (15.68–18.63 g/100 g) [6]. According to the ideal model proposed by FAO/WHO, the EAA/TAA ratio for high quality proteins should be around 40%, and *P. semisulcatus*'s EAA/TAA ratio generally meets this requirement, indicating a good amino acid balance. However, the overall amino acid level in *P. semisulcatus* is relatively low, which may suggest nutritional deficiencies or environmental stress. Among the three populations of *P. semisulcatus*, the PsHK population exhibited relatively higher amino acid levels, demonstrating an advantage in terms of amino acid content.

4.3. Comparative Analysis of Fatty Acid Profiles Across P. semisulcatus Populations

Fatty acids in aquatic organisms fulfill several essential functions. They are key components of cell membranes, where they influence membrane fluidity and functionality. Additionally, fatty acids serve as a primary energy source. Through β-oxidation in the mitochondria, fatty acids release energy that is critical for the growth, development, and reproduction of aquatic organisms [41]. Particularly, polyunsaturated fatty acids (PUFAs) regulate inflammatory responses, promote cell proliferation and differentiation, and are critical for the growth, development, reproductive health, and maintenance of the nervous system in aquatic organisms [42–44]. Monounsaturated fatty acids (MUFAs) have a positive effect on promoting blood circulation, regulating blood glucose and lipid metabolism, and effectively lowering cholesterol levels [45]. Additionally, certain fatty acids, such as eicosapentaenoic acid (EPA) and docosahexaenoic acid (DHA), have positive effects on the immune system of aquatic organisms, enhancing disease resistance and improving survival rates [46]. These fatty acids are converted in aquatic organisms through a series of dehydrogenation and elongation reactions, playing a key role in maintaining the integrity and function of cell membranes [47].

In this study, *P. semisulcatus* had the highest content of C18:0 and the lowest content of C20:2, a trend similar to that observed in M. ensis [30]. All samples showed a higher content of polyunsaturated fatty acids (ΣPUFA) compared to monounsaturated fatty acids (ΣMUFA). The saturated fatty acid content ranged from 176.70–222.50 mg/100 g, slightly higher than that of *P. monodon* (125.49–221.79 mg/100 g) [6]. The monounsaturated fatty acid content ranged from 59.10–91.27 mg/100 g, slightly lower than that in *M. ensis* (75.1–104.2 mg/100 g) [30] but higher than in *L. vannamei* (14.79–17.67 mg/100 g) [28] and *F. merguiensis* (18.85 mg/100 g) [33]. Moreover, the DHA + EPA content in this study ranged from 103.70–123.20 mg/100 g, higher than that of *F. merguiensis* (90.20–131.4 mg/100 g) [31] and *Penaeus japonicus* (11.58 mg/100 g) [32]. Overall, *P. semisulcatus* has relatively high levels of fatty acids and shows similar trends to other shrimp species, indicating a higher nutritional value and relatively balanced fatty acid levels. Among the three populations of *P. semisulcatus*, the PsHK population exhibited the highest levels of all fatty acids, suggesting that PsHK has a relatively higher nutritional value in terms of fatty acid content.

4.4. Comparative Analysis of Antioxidant Enzyme Activity Across P. semisulcatus Populations

Antioxidant enzymes in aquatic animals, such as T-AOC, SOD, and CAT, play crucial roles in the organism's defense system against oxidative stress. These enzymes work together to protect cells from damage caused by reactive oxygen species (ROS) [48,49]. T-AOC is a key indicator of antioxidant capacity, helping to neutralize free radicals and reactive oxygen species (ROS) in the body to prevent oxidative damage [50]. SOD primarily converts superoxide anion radicals into hydrogen peroxide and oxygen, thereby reducing the accumulation of free radicals and protecting cells from oxidative stress [48]. CAT further catalyzes the breakdown of hydrogen peroxide into water and oxygen, preventing its conversion into more harmful hydroxyl radicals in the presence of metal ions [51]. It is a key defense against the accumulation of hydrogen peroxide in cells. These antioxidant enzymes not only protect aquatic organisms from oxidative stress but also help maintain the redox balance within the body under environmental stress conditions, mitigate oxidative damage, and enhance adaptability and immune function.

In this study, the total antioxidant capacity (T-AOC) activity in *P. semisulcatus* from three different regions in China ranged from 2.67–2.68 μmol/g, higher than that of *P. clarkii* (2.57 U/mg) [52]. The SOD activity ranged from 50.07 to 54.96 μmol/(min·g), which is higher than that of *P. monodon* (39.65–47.29 U/g) [6] and *F. merguiensis* (46.85–53.98 U/g) [31]. Higher levels of T-AOC, SOD, and CAT indicate stronger antioxidant and immune capabilities. *P. semisulcatus* exhibits higher antioxidant enzyme activity compared to other shrimp species, suggesting it has stronger antioxidant and immune capabilities, enabling it to better resist oxidative stress. Among the three populations of *P. semisulcatus*, the PsHK group displayed the highest antioxidant enzyme activity, indicating that it has a stronger antioxidant capacity within these farming populations.

4.5. Genetic Diversity Comparison Among P. semisulcatus Populations

Genetic diversity is the basis of biological variation and a cornerstone for species' adaptation, survival, and evolutionary potential in complex and ever-changing environments [53]. Increased genetic diversity helps mitigate the detrimental effects of inbreeding, reduces the impact of genetic drift, and promotes the overall health and vitality of populations. Genetic diversity also underpins selective breeding in aquaculture, enabling the improvement of economically important traits like growth rate, meat quality, and disease resistance, thus enhancing farming efficiency. SNP density indicates the frequency of SNP loci in the genome, with higher density typically linked to increased genetic variation [54]. Nucleotide polymorphism (π) measures sequence variation within specific genomic regions and serves as a key parameter for evaluating intraspecific genetic diversity [55]. In the genetic diversity analysis of *P. semisulcatus*, the PsHK population showed higher values in various genetic parameters compared to the other two populations. The analytical results indicate that all three investigated *P. semisulcatus* populations exhibit comparatively limited genetic variability. Notably, the PsHK population displayed marginally elevated genetic diversity levels within this constrained variability framework.

5. Conclusions

This study systematically evaluated the nutritional composition, antioxidant properties, and genetic diversity of three cultured populations of *P. semisulcatus*. The results indicated that the PsHK population had relatively higher amino acid and fatty acid contents, reflecting superior nutritional value. Antioxidant enzyme activity showed no significant differences between populations, indicating similar antioxidant capacities. Genetic analysis revealed that the PsHK population exhibited the highest genetic diversity and greater genetic differentiation from the PsRP population. Overall, *P. semisulcatus* exhibited aquaculture traits of low protein and high fat, with good nutritional quality, making it suitable for broader aquaculture promotion. Among the three populations, PsHK stood out in both nutritional composition and genetic background, providing valuable insights for future germplasm conservation, genetic improvement, and industrial development.

Author Contributions: F.Z. and T.L. conceived the project and supervised the work. Y.L., Z.J. and J.C. performed the bioinformatics analysis and prepared the manuscript, the table and figures. Y.L. and J.C. conducted the experiment. L.Y., S.J., Q.Y., J.H., Y.D. and J.S. collected the samples and performed sequencing. All authors have read and agreed to the published version of the manuscript.

Funding: This research was funded by National Key R&D Program of China (2022YFD2401900); Central Public-interest Scientific Institution Basal Research Fund, CAFS (2023TD34); China Agriculture Research System (CARS-48); Guangdong Basic and Applied Basic Research Foundation (2023A1515012410); Hainan Provincial Natural Science Foundation of China (323MS127); Earmarked fund for HNARS (HNARS-10-ZJ01); Provincial Rural Revitalization Strategy Special Fund Seed Industry Revitalization Project (2024-SPY-00-005); Hainan Seed Industry Laboratory-Joint Unveiling of the Project (B24YQ0010); Central Public-interest Scientific Institution Basal Research Fund, South China Sea Fisheries Research Institute, CAFS (No. 2024RC06); Research on breeding technology of candidate species for Guangdong modern marine ranching (2024-MRB-00-001).

Institutional Review Board Statement: The animal study was reviewed and approved by the Animal Care and Use Committee of the South China Sea Fisheries Research Institute.

Data Availability Statement: The original contributions presented in this study are included in the article. Further inquiries can be directed to the corresponding authors.

Conflicts of Interest: The authors declare no conflicts of interest.

References

1. Halim, S.A.A.A.; Othman, A.S.; Akib, N.A.M.; Jamaludin, N.-A.; Esa, Y.; Nor, S.A.M. Mitochondrial Markers Identify a Genetic Boundary of the Green Tiger Prawn (*Penaeus semisulcatus*) in the Indo-Pacific Ocean. *Zool. Stud.* **2021**, *60*, 8. [CrossRef]
2. Kumlu, M.; Eroldogan, O.T.; Aktas, M. Effects of Temperature and Salinity on Larval Growth, Survival and Development of *Penaeus semisulcatus*. *Aquaculture* **2000**, *188*, 167–173. [CrossRef]
3. Rajkumar, M.; Pillai, S.L.; Rameshkumar, P.; Saravanan, R.; Thirumalaiselvan, S.; Jose, J.; Sobhana, K.S.; George, M.R. Reproductive Biology of Green Tiger Shrimp *Penaeus semisulcatus* De Haan, 1844 in Palk Bay, Southeast Coast of India. *Reg. Stud. Mar. Sci.* **2023**, *66*, 103161.
4. Niamaimandi, N.; Aziz, A.; Siti Khalijah, D.; Che Roos, S.; Kiabi, B. Reproductive Biology of the Green Tiger Prawn (*Penaeus semisulcatus*) in Coastal Waters of Bushehr, Persian Gulf. *ICES J. Mar. Sci.* **2008**, *65*, 1593–1599.
5. Alrashada, Y.N.; Tharwat, A.; Boqursain, A. Population Dynamics of the Green Tiger Prawn, *Penaeus semisulcatus* de Haan, 1844 in the Saudi Coast of the Arabian Gulf. *Indian J. Anim. Res.* **2019**, 1–6. [CrossRef]
6. Li, Y.; Chen, J.; Jiang, S.; Huang, J.; Jiang, S.; Yang, Q.; Yang, L.; Shi, J.; Zhou, F. A Comprehensive Study on Nutritional Quality, Physiological Enzyme Activity and Genetic Diversity in Six Populations of Penaeus Monodon. *Aquacult. Int.* **2024**, *32*, 10141–10157. [CrossRef]
7. Li, Y.; Cao, S.; Jiang, S.; Huang, J.; Yang, Q.; Jiang, S.; Yang, L.; Zhou, F. Comparative Study of Nutritional Composition, Physiological Indicators, and Genetic Diversity in *Litopenaeus vannamei* from Different Aquaculture Populations. *Biology* **2024**, *13*, 722. [CrossRef]
8. Meng, X.; Fu, Q.; Luan, S.; Luo, K.; Sui, J.; Kong, J. Genome Survey and High-Resolution Genetic Map Provide Valuable Genetic Resources for *Fenneropenaeus chinensis*. *Sci. Rep.* **2021**, *11*, 7533.
9. El-Gendy, A.M.; El-Feky, F.; Mahmoud, N.H.; Elsebakhy, G.S. Evaluation of Nutritional Quality of Green Tiger Prawn, *Penaeus semisulcatus* from Land Fisheries (Alexandria) and Market (India). *Egypt. J. Hosp. Med.* **2018**, *70*, 924–934. [CrossRef]
10. Tamadoni Jahromi, S.; Sofman Othman, A.; Rosazlina, R.; Pourmozaffar, S.; Gozari, M. Population Genetics of *Penaeus semisulcatus* from Persian Gulf and Oman Sea Using Newly Developed DNA Microsatellite Markers. *Iran. J. Fish. Sci.* **2021**, *20*, 157–178. [CrossRef]
11. Shuhua, T.A.N.; Guizhong, W.; Qiongwu, L.I.N. Genetic Diversity of Two Wild Populations of *Penaeus semisulcatus* Revealed by RAPD Technique. *Acta Ecol. Sin.* **2006**, *26*, 3907–3911.
12. *GB 5009.3-2016*; Determination of Moisture in Food. National Standards of the People's Republic of China: Beijing, China, 2016.
13. *GB 5009.4-2016*; National Food Safety Standard—Determination of Ash in Food. National Standards of the People's Republic of China: Beijing, China, 2016.
14. *GB 5009.5-2016*; National Standards for Food Safety—Determination of Proteins in Foods. National Standards of the People's Republic of China: Beijing, China, 2016.
15. *GB 5009.6-2016*; National Food Safety Standard—Determination of Crude Fat in Food. National Standards of the People's Republic of China: Beijing, China, 2016.
16. *GB/T 15672-2009*; Determination of Total Sugar Content in Edible Fungi. National Standards of the People's Republic of China: Beijing, China, 2009.
17. *GB 5009.124-2016*; National Food Safety Standard—Determination of Amino Acids in Foods by Hydrochloric Acid Hydrolysis. National Standards of the People's Republic of China: Beijing, China, 2016.
18. *GB 5009.168-2016*; National Food Safety Standard—Determination of Fatty Acids in Foods by Gas Chromatography. National Standards of the People's Republic of China: Beijing, China, 2016.
19. Jin, X.; Wang, X.; Tse, W.K.F.; Shi, Y. Homeostasis and Physiological Regulation in the Aquatic Animal during Osmotic Stress. *Front. Physiol.* **2022**, *13*, 977185. [CrossRef]

20. Boyd, C.E.; Lawrence, J.M. The Mineral Composition of Several Fresh-Water Algae. 1967. Available online: https://seafwa.org/sites/default/files/journal-articles/BOYD-413.pdf (accessed on 1 March 2025).
21. Lall, S.P.; Kaushik, S.J. Nutrition and Metabolism of Minerals in Fish. *Animals* **2021**, *11*, 2711. [CrossRef] [PubMed]
22. Sun, Z. Foundations for the Study of Structure and Function of Proteins. In *Basics of Bioinformatics*; Jiang, R., Zhang, X., Zhang, M.Q., Eds.; Springer: Berlin/Heidelberg, Germany, 2013; pp. 303–336; ISBN 978-3-642-38950-4.
23. Alhotan, R.A.; Pesti, G.M.; Billard, L. The Linear Relationship between True Protein and Nitrogen Contents of Feed and Food Ingredients: Calculating True Protein from a New Conversion Factor. *Cogent Food Agric.* **2024**, *10*, 2428821. [CrossRef]
24. Ahmed, I.; Jan, K.; Fatma, S.; Dawood, M.A.O. Muscle Proximate Composition of Various Food Fish Species and Their Nutritional Significance: A Review. *Anim. Physiol. Nutr.* **2022**, *106*, 690–719. [CrossRef] [PubMed]
25. Kainz, M.; Brett, M.T.; Arts, M.T. (Eds.) *Lipids in Aquatic Ecosystems*; Springer: New York, NY, USA, 2009; ISBN 978-0-387-88607-7.
26. dos Santos Carvalho, C.; Fernandes, M.N. Effect of Copper on Liver Key Enzymes of Anaerobic Glucose Metabolism from Freshwater Tropical Fish *Prochilodus lineatus*. *Comp. Biochem. Physiol. Part A Mol. Integr. Physiol.* **2008**, *151*, 437–442. [CrossRef]
27. Dos Santos Carvalho, C.; Fernandes, M.N. Effects of Copper Toxicity at Different pH and Temperatures on the in Vitro Enzyme Activity in Blood and Liver of Fish, *Prochilodus lineatus*. *Mol. Biol. Rep.* **2019**, *46*, 4933–4942. [CrossRef]
28. Li, X.; Wang, Y.; Li, H.; Jiang, X.; Ji, L.; Liu, T.; Sun, Y. Chemical and Quality Evaluation of Pacific White Shrimp *Litopenaeus vannamei*: Influence of Strains on Flesh Nutrition. *Food Sci. Nutr.* **2021**, *9*, 5352–5360. [CrossRef]
29. Xu, L.; Yan, B. Nutritional Component Analysis and Quality Evaluation of Penaeus Japonicus. *Food Sci.* **2011**, *32*, 297–301.
30. Li, Y.; Chen, J.; Jiang, S.; Yang, Q.; Yang, L.; Huang, J.; Shi, J.; Zhang, Y.; Lu, Z.; Zhou, F. A Comprehensive Assessment of Nutritional Value, Antioxidant Potential, and Genetic Diversity in *Metapenaeus ensis* from Three Different Populations. *Biology* **2024**, *13*, 838. [CrossRef]
31. Li, Y.; Chen, J.; Cao, S.; Jiang, Z.; Jiang, S.; Yang, Q.; Yang, L.; Huang, J.; Shi, J.; Ma, Z. A Comprehensive Assessment of the Nutritional Value, Antioxidant Potential, and Genetic Diversity of *Fenneropenaeus merguiensis* from Three Different Regions in China. *Biology* **2024**, *13*, 1002. [CrossRef] [PubMed]
32. Liu, Z.; Liu, Q.; Zhang, D.; Wei, S.; Sun, Q.; Xia, Q.; Shi, W.; Ji, H.; Liu, S. Comparison of the Proximate Composition and Nutritional Profile of Byproducts and Edible Parts of Five Species of Shrimp. *Foods* **2021**, *10*, 2603. [CrossRef]
33. Wang, J. Comparision of Nutritional Composition in Muscle of *Penaeus chinensis*, *Penaeus vannamei* Boone and *Penaeus japonicuss* Bate. *Food Sci. Technol.* **2013**, *38*, 145–150.
34. Li, P.; Yin, Y.-L.; Li, D.; Kim, S.W.; Wu, G. Amino Acids and Immune Function. *Br. J. Nutr.* **2007**, *98*, 237–252.
35. Huang, M.; Gao, Q.; Yang, X.; Jiang, W.; Hao, L.; Yu, Y.; Tian, Y. Free Amino Acids in Response to Salinity Changes in Fishes: Relationships to Osmoregulation. *Fish Physiol. Biochem.* **2023**, *49*, 1031–1042. [CrossRef] [PubMed]
36. Wu, G. Functional Amino Acids in Nutrition and Health. *Amino Acids* **2013**, *45*, 407–411. [CrossRef]
37. Niu, J.; Hu, X.L.; Ip, J.C.; Ma, K.Y.; Tang, Y.; Wang, Y.; Qin, J.; Qiu, J.-W.; Chan, T.F.; Chu, K.H. Multi-Omic Approach Provides Insights into Osmoregulation and Osmoconformation of the Crab *Scylla paramamosain*. *Sci. Rep.* **2020**, *10*, 21771. [PubMed]
38. Huo, D.; Zhang, L.; Yang, H.; Sun, L. Adaptation to Hypoxic Stress Involves Amino Acid Metabolism: A Case in Sea Cucumber. *Environ. Pollut.* **2023**, *330*, 121766. [CrossRef]
39. Batista-Silva, W.; Heinemann, B.; Rugen, N.; Nunes-Nesi, A.; Araújo, W.L.; Braun, H.; Hildebrandt, T.M. The Role of Amino Acid Metabolism During Abiotic Stress Release. *Plant Cell Environ.* **2019**, *42*, 1630–1644. [CrossRef]
40. Zhu, J.; Xu, H.; Zou, Z.; Yao, H.; Lin, Z.; Dong, Y. Effects of Medium-and Long-Term High-Salinity Environments on Free Amino Acid Content and Related Genes of *Sinonovacula constricta*. *Front. Mar. Sci.* **2024**, *11*, 1368952. [CrossRef]
41. Houten, S.M.; Violante, S.; Ventura, F.V.; Wanders, R.J.A. The Biochemistry and Physiology of Mitochondrial Fatty Acid β-Oxidation and Its Genetic Disorders. *Annu. Rev. Physiol.* **2016**, *78*, 23–44. [CrossRef] [PubMed]
42. Kang, J.X.; Weylandt, K.H. Modulation of Inflammatory Cytokines by Omega-3 Fatty Acids. In *Lipids in Health and Disease*; Quinn, P.J., Wang, X., Eds.; Subcellular Biochemistry; Springer: Dordrecht, The Netherlands, 2008; Volume 49, pp. 133–143; ISBN 978-1-4020-8830-8.
43. Fritsche, K.L. The Science of Fatty Acids and Inflammation. *Adv. Nutr.* **2015**, *6*, 293S–301S. [PubMed]
44. Rashid, M.A.; Haque, M.; Akbar, M. Role of Polyunsaturated Fatty Acids and Their Metabolites on Stem Cell Proliferation and Differentiation. In *The Benefits of Natural Products for Neurodegenerative Diseases*; Essa, M.M., Akbar, M., Guillemin, G., Eds.; Advances in Neurobiology; Springer International Publishing: Cham, Switzerland, 2016; Volume 12, pp. 367–380; ISBN 978-3-319-28381-4.
45. Cao, X.; Xia, J.; Zhou, Y.; Wang, Y.; Xia, H.; Wang, S.; Liao, W.; Sun, G. The Effect of Mufa-Rich Food on Lipid Profile: A Meta-Analysis of Randomized and Controlled-Feeding Trials. *Foods* **2022**, *11*, 1982. [CrossRef]
46. Calder, P.C. N-3 Fatty Acids, Inflammation and Immunity: New Mechanisms to Explain Old Actions. *Proc. Nutr. Soc.* **2013**, *72*, 326–336. [PubMed]
47. Lee-Okada, H.-C.; Xue, C.; Yokomizo, T. Recent Advances on the Physiological and Pathophysiological Roles of Polyunsaturated Fatty Acids and Their Biosynthetic Pathway. *Biochim. Biophys. Acta (BBA)-Mol. Cell Biol. Lipids* **2025**, *1870*, 159564.

48. Zhang, L.; Zhao, M.; Wang, Y. A Review on Superoxide Dismutases of Hydrobios. 2012; pp. 800–801. Available online: https://www.cabidigitallibrary.org/doi/full/10.5555/20123388727 (accessed on 1 March 2025).
49. Attia, H.G.; El-Morshedy, S.M.; Nagy, A.M.; Ibrahim, A.M.; Aleraky, M.; Abdelrahman, S.S.; Osman, S.M.; Alasmari, S.M.; El Raey, M.A.; Abdelhameed, M.F. Citrus Clementine Peel Essential Oil Ameliorates Potassium Dichromate-Induced Lung Injury: Insights into the PI3K/AKT Pathway. *Metabolites* **2024**, *14*, 68. [CrossRef]
50. Silvestrini, A.; Meucci, E.; Ricerca, B.M.; Mancini, A. Total Antioxidant Capacity: Biochemical Aspects and Clinical Significance. *Int. J. Mol. Sci.* **2023**, *24*, 10978. [CrossRef]
51. Carmo De Carvalho E Martins, M.D.; Martins; Da Silva Santos Oliveira, A.S.; Da Silva, L.A.A.; Primo, M.G.S.; De Carvalho Lira, V.B. Biological Indicators of Oxidative Stress [Malondialdehyde, Catalase, Glutathione Peroxidase, and Superoxide Dismutase] and Their Application in Nutrition. In *Biomarkers in Nutrition*; Patel, V.B., Preedy, V.R., Eds.; Biomarkers in Disease: Methods, Discoveries and Applications; Springer International Publishing: Cham, Switzerland, 2022; pp. 1–25; ISBN 978-3-030-81304-8.
52. Tan, S.H.; Deng, X.Y.; Jiang, W.M.; He, F.L. Effects of High Level Chromium on Antioxidant Enzyme System in Gill and Hepatopancreas of Procambarus Clarkii. *J. Agro-Environ. Sci.* **2007**, *23*, 24–34.
53. O'Connell, M.; Wright, J.M. Microsatellite DNA in Fishes. *Rev. Fish Biol. Fish.* **1997**, *7*, 331–363. [CrossRef]
54. Matukumalli, L.K.; Lawley, C.T.; Schnabel, R.D.; Taylor, J.F.; Allan, M.F.; Heaton, M.P.; O'Connell, J.; Moore, S.S.; Smith, T.P.; Sonstegard, T.S. Development and Characterization of a High Density SNP Genotyping Assay for Cattle. *PLoS ONE* **2009**, *4*, e5350.
55. Luo, W.; Luo, C.; Wang, M.; Guo, L.; Chen, X.; Li, Z.; Zheng, M.; Folaniyi, B.S.; Luo, W.; Shu, D. Genome Diversity of Chinese Indigenous Chicken and the Selective Signatures in Chinese Gamecock Chicken. *Sci. Rep.* **2020**, *10*, 14532. [PubMed]

Journal of
*Marine Science
and Engineering*

MDPI

Article

Adaptive Responses in Byssal Growth and Shedding: Insights from *Pteria penguin* Under Thread Trimming and Non-Trimming Conditions

Hebert Ely Vasquez [1,†], Shangkun Wei [1,2,†], Guoliang Yang [1,2], Lingfeng Wang [1,2], Peixuan Yu [1,2], Mingyue Dong [1,2], Chao Yuan [3,*] and Xing Zheng [1,2,4,*]

1 School of Marine Biology and Fisheries, Hainan University, Haikou 570228, China; hebertely@163.com (H.E.V.); w103077246999@163.com (S.W.); yanggl@hainanu.edu.cn (G.Y.); 19951957209@163.com (L.W.); 23210707030014@hainanu.edu.cn (P.Y.); 13166537229@163.com (M.D.)
2 Hainan Key Laboratory of Tropical Hydrobiological Technology, Hainan University, Haikou 570228, China
3 Haikou Marine Center of the Ministry of Natural Resources, Haikou 570100, China
4 Sanya Tropical Fisheries Research Institute, Sanya 572018, China
* Correspondence: 18689761212@163.com (C.Y.); zhengxing_edu@163.com (X.Z.)
† These authors contributed equally to this work.

Abstract: Bivalves use byssal threads for attachment and locomotion, periodically shedding and regenerating them. In the winged pearl oyster *Pteria penguin*—known for its strong byssus and its role in the pearl industry—shedding may occur when the byssal stalk reaches a critical size, although the underlying mechanism remains unclear. This study investigated whether artificial manipulation of the byssus (via trimming) could stimulate thread production and promote shedding in adult *P. penguin* from two size groups. Byssal threads attached to the substrate were trimmed every 3–5 days over a 30-day period and compared to untrimmed controls. Oysters with trimmed byssus produced significantly more threads, with smaller individuals outperforming larger ones in both thread count and byssal stalk diameter. Moreover, small oysters exhibited a higher frequency of complete byssal shedding. These findings suggest that trimming stimulates thread production and accelerates stalk thickening, potentially triggering shedding due to spatial constraints at the attachment site. This response appears to reflect an adaptive mechanism for maintaining effective attachment and may help explain how mechanical or environmental cues influence byssal dynamics. Understanding this process offers new insight into the behavioral and physiological plasticity of *P. penguin*, with potential applications in pearl oyster management and aquaculture.

Keywords: *Pearl oyster*; *Pteria penguin*; thread trimming; byssal regrowth; stalk release

1. Introduction

Bivalves that produce byssus have evolved a wide variety of byssal structures, which play a crucial role in their ability to anchor themselves to intertidal and underwater substrates. These specialized attachments enable bivalves to maintain stability in dynamic aquatic environments, contributing significantly to their survival, ecological, and evolutionary success [1–3]. This versatility is also manifested in pearl oysters, which depend on these strong, fibrous threads to anchor themselves to various surfaces. However, the strategies for managing these attachments can differ considerably between species [4]. For example, *Pinctada margaritifera* and *P. fucata* are known to continually produce byssus throughout their lives, ensuring they remain firmly anchored in their benthic environment [5]. In contrast, *P. maxima* stops secreting byssus once it grows large and heavy enough to stay in place despite the challenges posed by ocean currents [6]. The growth of byssal threads in these oysters is not just a matter of species; it is also influenced by a range of environmental factors like water temperature [7], pH levels [8,9], salinity [10,11], and the strength of local currents [6,12]. The effect of these factors on the byssogenesis in pearl oysters is evident, and as oysters age, their production of byssal threads tends to decrease considerably [6,13]. But as the byssal threads are constantly produced, they are also being released or shed, and this is a phenomenon that has not been explored in depth in other bivalves as in mussels [14]. Pearl oysters can also shed their byssi; they do this by discarding the old threads and then secreting new ones, sometimes even before the old threads have completely detached [13,15]. Byssal shedding allows oysters to engage in active displacement, which can be crucial for survival in challenging environmental conditions such as high predation risk or extreme environmental stressors, including temperature fluctuations and salinity changes [16]. Similar to other byssally attached species, the dynamics of byssal production, attachment, and release can be used as a form of locomotion by short movements [17].

The winged pearl oyster *Pteria penguin* belongs to the Pteriidae family, a byssally attached bivalve that is not considered a "true" oyster according to malacologists [18] and is well known for their use in the production of half pearls (mabe) [19,20]. There have been difficulties in attempts to produce round pearls due to the low nuclei retention rate. This is a problem that has been linked to a large byssus, as in other pearl oysters [21]. Excessive byssal development may interfere with the success of the grafting procedure by causing stress or limiting the available energy for pearl sac development, ultimately reducing pearl quality or retention. Understanding and potentially managing byssal production could therefore contribute to improved grafting success and overall pearl yield in aquaculture settings. After metamorphosis, *P. penguin* juveniles retain the byssal glands producing byssal threads throughout their adult life, as in many pterioideans [22]. Interestingly, despite possessing one of the strongest byssi among bivalves, *P. penguin* can also shed its byssus completely, including the byssal root, and young adult individuals raised in lantern nets are occasionally found without byssus [23]. Moreover, the length and diameter of the byssal stalk are more closely tied to the oyster's weight than to the size of its shell [23], and as the oysters continue to gain weight, there comes a point where the diameter of their byssal stalks ceases to increase [13]. This observation suggests the presence of a physiological threshold, beyond which the byssus can no longer thicken, regardless of further growth of the oyster.

To further explore this phenomenon, this study aims to investigate the impact of stimulating byssal growth and shedding by periodically detaching the oyster from its substrate through the mechanical trimming of its byssal threads. The underlying hypothesis is that by regularly trimming the newly produced byssal threads, the oysters will respond by producing new byssal threads that ultimately will enlarge the byssus, leading to the voluntary shedding of the whole byssal organ. This approach aims to provide deeper insights into the adaptive mechanisms governing byssal production and shedding in pearl oysters.

2. Materials and Methods

2.1. Experimental Animals and Conditions

A total of 100 adult *P. penguin* oysters were collected from an oyster farm in Sanya, Hainan Province (China), where the oysters were cultured in semi-natural open-ocean conditions, characterized by high water quality and a suitable environment for suspended culture using lantern nets hung from rafts. The oysters were transported in wet conditions inside styrofoam boxes to the Sanya Tropical Fisheries Research Institute (Lingshui,

Hainan Province). These oysters were divided into two distinct size groups: small oysters (65.11 ± 7.07 mm maximum dorsoventral length, MDL; 10.46 ± 3.61 g wet weight, WW) and large oysters (87.34 ± 8.03 mm MDL, 24.24 ± 5.66 g WW), with each group comprising 50 individuals. Although likely from the same 10-month-old cohort, natural variability in growth resulted in these two size classes, which were used to assess potential size-related responses. Each individual was tagged using waterproof paper labels (10 mm × 10 mm), which were affixed to the oyster shell with a drop of fast-drying cyanoacrylate (super glue) to ensure durable identification throughout the experiment. After the adhesive had fully dried—typically within 5 min—the oysters were thoroughly rinsed with seawater and placed into flat net cages, each capable of accommodating 25 oysters. Inside the cages, oysters were vertically arranged in a manner that allowed each oyster's foot to reach the adjacent oyster shell, thereby promoting byssal attachment (Figure 1). The experiment was conducted over a 30-day period, allowing for consistent monitoring of byssal production and shedding behaviors across individuals.

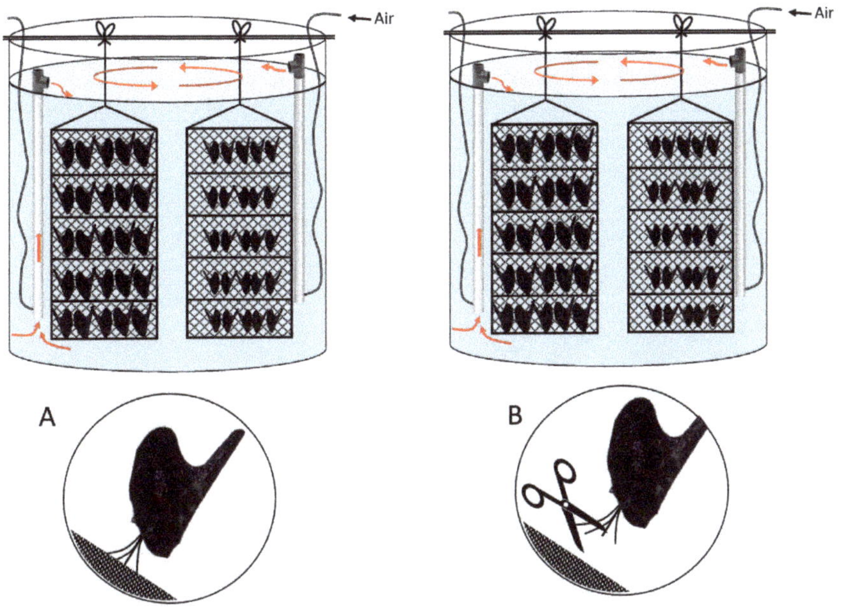

Figure 1. Experimental setup for byssal production and shedding. Red arrows indicate the direction of the water movement in the tanks. (**A**) Oyster with untrimmed byssus; (**B**) oyster with trimmed byssus (every 3–5 days).

All oysters were allowed acclimation for 13 days in 200 L tanks under constant controlled husbandry conditions during this period and throughout this study. These conditions included daily 100% water changes, ad libitum feeding with a mixed diatom diet (*Skeletonema* sp. and *Chaetoceros muelleri*), a water temperature of 28 ± 1 °C using filtered seawater (10-micron polypropylene felt filter bag), and salinity levels maintained at 32 PSU. Aeration through upwelling was employed to create a circular water flow in the tanks, ensuring optimal oxygenation and water movement.

2.2. Byssal Thread Counts, Diameter Measure, and Trimming

To observe byssal thread production and shedding, two experimental conditions were established. In the first condition, a cage containing 25 small oysters and another cage containing 25 large oysters were submerged together in one tank. No byssal trimming was performed, allowing for observation of natural byssal shedding (Figure 1A). In the second condition, another two cages—one with small oysters and the other with large oysters—were similarly prepared and submerged in a separate tank. Throughout the experimental period, byssal threads were periodically counted and trimmed using scissors to simulate byssal shedding and observe the oysters' responses to this intervention (Figure 1B). By maintaining these two distinct conditions, the experiment aimed to provide a comprehensive understanding of how byssal production and shedding are influenced by size and external manipulation in *P. penguin*. Trimmed and untrimmed oysters were maintained in separate tanks without within-treatment replication, due to logistical and space constraints. While this design enabled clear monitoring of treatment effects, potential tank effects and associated bias cannot be fully excluded.

The number of single byssal threads secreted by each oyster was counted every 3–5 days over 30 days. To standardize the measurements, the number of threads was counted at the distal section of the byssus because threads suffered fusion as threads got closer to the proximal section of the byssus. The absolute number of byssal threads produced by each oyster was obtained by subtracting the thread count from the immediate previous count.

The diameter of the byssus in all oysters was measured at the commencement (day 0) and the end of the experiment (day 30) by opening the oyster valves using a pair of pearl oyster openers. The measurements of the byssal diameter (BD) at the proximal section were acquired using a vernier caliper (0.1 mm). The trimming of the byssus was carried out only on oysters that were found attached and was carried out every 3–5 days by using a pair of scissors (Figure 1B). The equation (1) used for calculating the change in byssal diameter (Δ BD) was

$$\Delta \text{ BD} = (D_1 - D_0) \quad (1)$$

where Δ BD represents the change in diameter, D_0 represents the diameter at day 0, and D_1 represents the diameter at day 30.

2.3. Data Analysis

A violin plot was constructed using the R package 'ggstatplot' [24]. Welch's ANOVA, followed by the Games–Howell test for pairwise comparisons, was used to compare the absolute count of byssal threads among the four experimental groups with the function 'ggbetweenstats'. These robust parametric methods were chosen because they do not assume equal variances or balanced sample sizes, which were conditions not fully met by our data.

Boxplots were constructed using the package 'ggplot2' [25]. To assess the differences between the accumulated number of byssal threads produced for each oyster at any given sampled day, the 'lm()' function was used to fit a linear model with the absolute number of threads as the dependent variable and day as the independent variable. The 'emmeans()' function was then applied to this linear model object to compute estimated marginal means and conduct a Tukey post hoc test.

The association degree between the variables ΔBD and the total number of threads produced by the oysters whose byssi were trimmed was calculated by the determination coefficient (R^2) after the data underwent handling of outliers and log transformation, and its level of significance was determined by fitting the data using the 'lm()' function. To assess the goodness of fit and assumptions of the linear regression model, a diagnostic plot was generated using the 'fitPlot()' function from the 'FSAmisc' package [26]. All differences were considered significant at $p < 0.05$. Plotting, curve fitting, and statistical analyses were carried out using R [27].

3. Results

3.1. Survival Rate and Thread Counts in Oysters with Trimmed and Untrimmed Byssus

The survival rate of oysters in most experimental groups was 100%, except for the group of large oysters whose byssus remained intact, which showed a survival rate of 90%. The absolute number of byssal threads consistently revealed that byssal thread production was significantly higher in oysters subjected to periodic trimming, regardless of their shell size, than in oysters with untrimmed byssus ($p < 0.05$, Figure 2). Oysters with untrimmed byssal threads displayed an average of 1.24 ± 2.48 threads in small individuals and 0.68 ± 2.74 threads in large ones. In contrast, oysters subjected to regular byssal trimming showed significantly higher averages, with small oysters producing 6.09 ± 3.68 threads and large oysters producing 3.96 ± 3.54 threads.

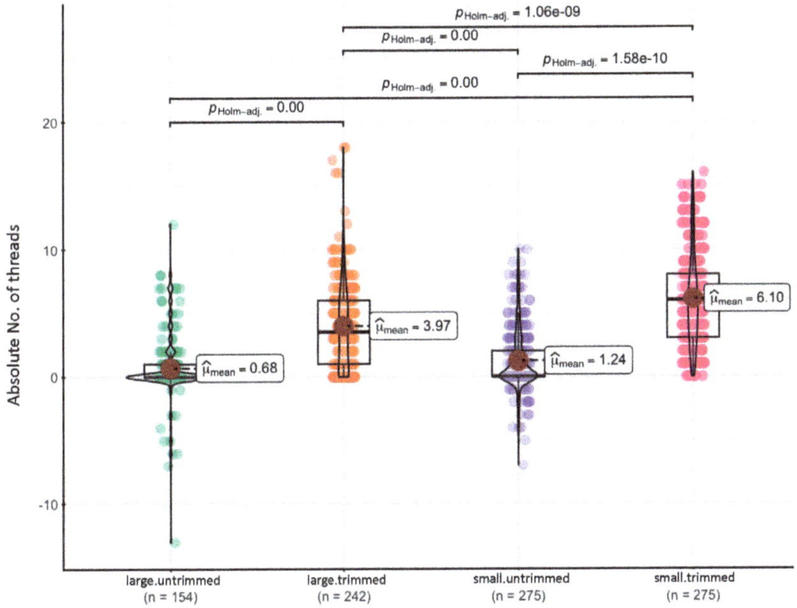

Figure 2. Absolute counts of byssal threads in small- and large-sized *P. penguin* oysters with trimmed and untrimmed byssi. Horizontal lines indicate significant differences in Games–Howell pairwise comparisons (F (3, 473.22) = 146.98, $p < 0.05$; n = 154–275). Negative values represent a net reduction in thread count in untrimmed oysters due to the fusion of previously distinct byssal threads, particularly toward the end of the experiment, where individual threads merged into thicker stalks.

The mean cumulative number of byssal threads produced by pearl oysters with trimmed and untrimmed byssal threads across size groups is shown in Figure 3. Oysters subjected to byssal trimming exhibited a consistently increasing trend in thread production over the 30-day observation period. Small and large oysters produced an average of 67.08 ± 23.31 and 43.63 ± 25.15 byssal threads, respectively, after their byssus was trimmed up to 10 times at regular intervals. The mean cumulative thread counts increased significantly from day 0 to day 30 in both size groups. By the end of the experiment, oysters with trimmed byssus displayed a wide range of cumulative thread counts, with the median value approaching the upper quartile, indicating robust thread production. Thread production was consistently high and significantly different from earlier time points, particularly after day 10 (Tukey HSD test, $p < 0.05$).

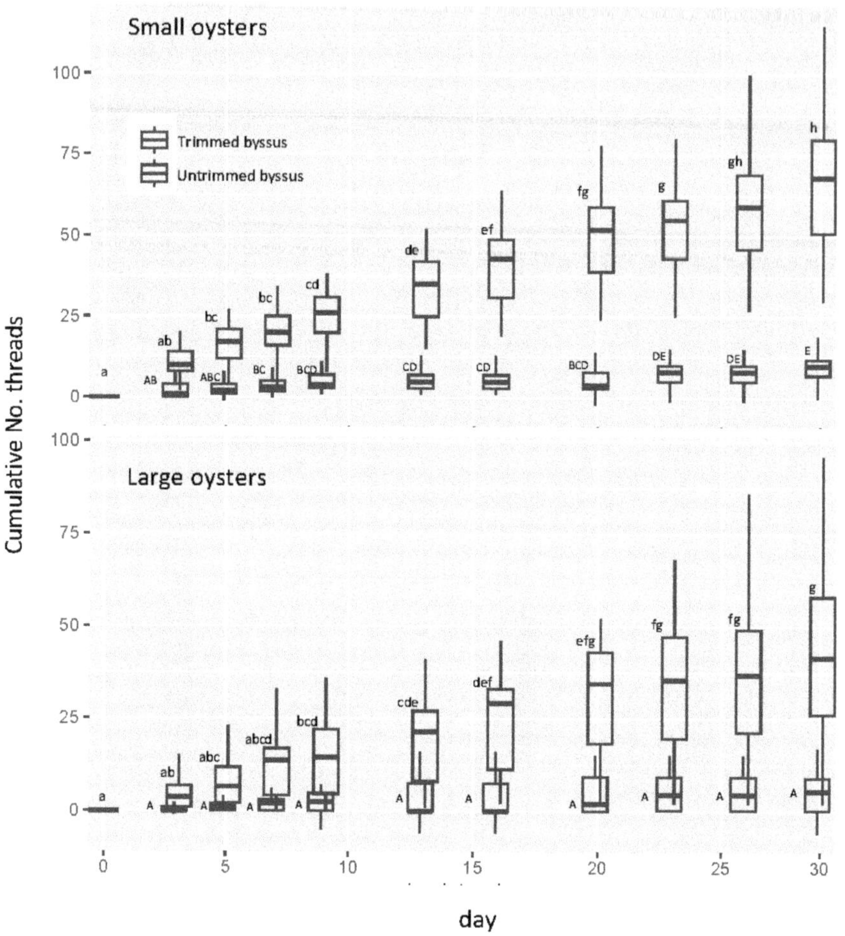

Figure 3. Mean cumulative number of byssal threads produced over 30 days in trimmed and untrimmed *P. penguin* oysters, grouped by size class (small and large individuals). "Trimmed byssus" refers to oysters whose byssal threads were periodically cut near the shell using scissors to simulate byssal shedding and encourage continued thread production. "Untrimmed byssus" refers to oysters whose byssal threads were left intact to observe natural patterns of thread production and shedding. Uppercase and lowercase letters indicate results of post hoc Tukey's HSD multiple comparisons among sampling days within each size group.

In contrast, oysters with untrimmed byssal threads produced far fewer threads. Small and large oysters in this group generated an average of 9.2 $\pm$ 5.2 and 5.28 $\pm$ 6.7 byssal threads, respectively. The variability in thread counts was lower in the small-oyster-size group, and thread production plateaued much earlier compared to the trimmed group (Figure 3). For the large-oyster-size group, there was no significant change in thread production throughout the experiment (Figure 3).

3.2. Byssal Diameter in Small and Large Oysters

The byssal diameters of both small and large oysters, with trimmed and untrimmed threads, are shown in Figure 4. Measurements were taken at two time points—day 0 and day 30—across all experimental groups. For small oysters, the initial mean byssal diameter on day 0 was 2.29 $\pm$ 1.17 mm for the untrimmed group and 2.75 $\pm$ 1.26 mm for the trimmed group. By day 30, the mean diameter increased to 3.62 ± 1.10 mm in the untrimmed group and 5.14 ± 1.81 mm in the trimmed group.

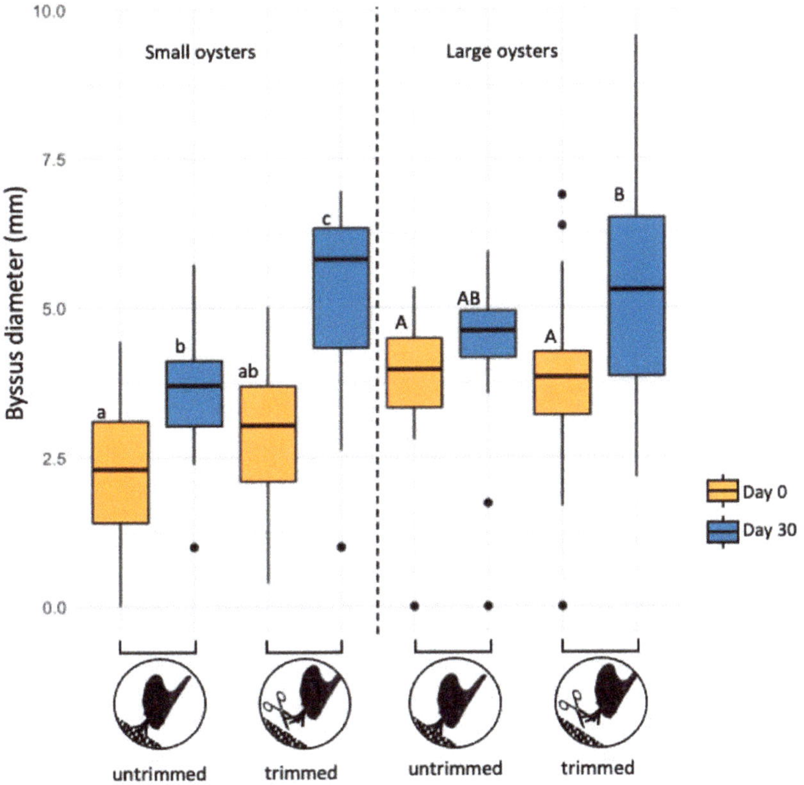

Figure 4. Boxplot showing the byssal diameters in small and large *P. penguin* oysters with trimmed and untrimmed byssus. Each box is composed of five percentiles that display the 10th, 25th, 50th, 75th, and 90th percentiles of a variable (the rectangle includes the 25th to 75th percentiles). All values above the 90th percentile and below the 10th percentile are plotted separately as black circles. Lowercase and uppercase letters indicate the result of post hoc HD Tukey test multiple comparisons among groups with small and large oysters, respectively ($n = 50$).

Similarly, for large oysters, the mean byssal diameter at day 0 was 3.77 ± 1.35 mm for the trimmed group and 3.76 ± 1.54 mm for the untrimmed group. By day 30, the mean diameter increased to 4.18 ± 1.57 mm for the untrimmed group and 5.29 ± 1.79 mm for the trimmed group. These findings indicate that both small and large oysters experienced an increase in byssal diameter over the 30-day experimental period. Furthermore, trimming the byssus appeared to stimulate greater byssal growth compared to the untrimmed byssus in both size categories.

A positive and statistically significant correlation was found between the total number of byssal threads and the change in byssal stalk diameter (ΔBD, Figure 5), with a coefficient of determination of $R^2 = 0.19$ ($p = 0.01$, $F = 6.55$).

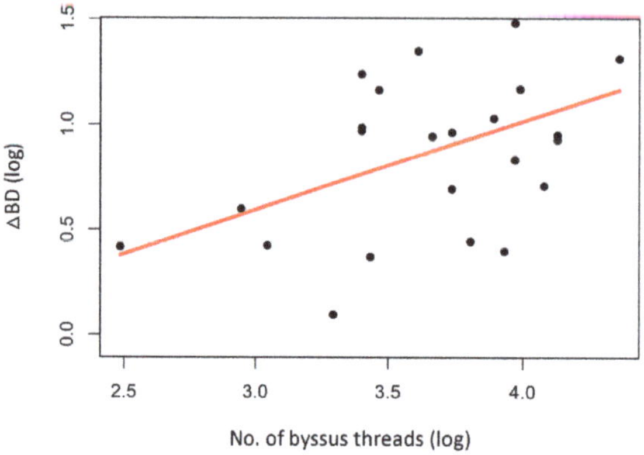

Figure 5. Correlation plot of the total number of byssal threads vs. byssal diameter in the pearl oyster *P. penguin* ($R^2 = 0.19$, $p = 0.01$, $F = 6.55$). Data were log-transformed and outliers were removed prior to analysis. Each point represents an individual oyster, with the x-axis showing the total number of byssal threads produced over the experimental period and the y-axis representing the change in mean byssal diameter. The solid line depicts the best-fit linear regression model generated using the lm() function.

3.3. Byssal Shedding in Small and Large Oysters

Small oysters with trimmed byssal threads exhibited a higher propensity for byssal shedding than those with intact byssus. By day 30, 36.0% of the small oysters subjected to byssal trimming had expelled their byssal threads, whereas only 4.1% of those with intact byssus exhibited such behavior (Figure 6A). In the large-oyster cohort, 22.7% showed byssal shedding when the threads were trimmed, while 14.3% did so when their byssus remained intact (Figure 6B). When oysters expelled (shed) their byssus, they did it completely, including the byssal root (Figure 6C).

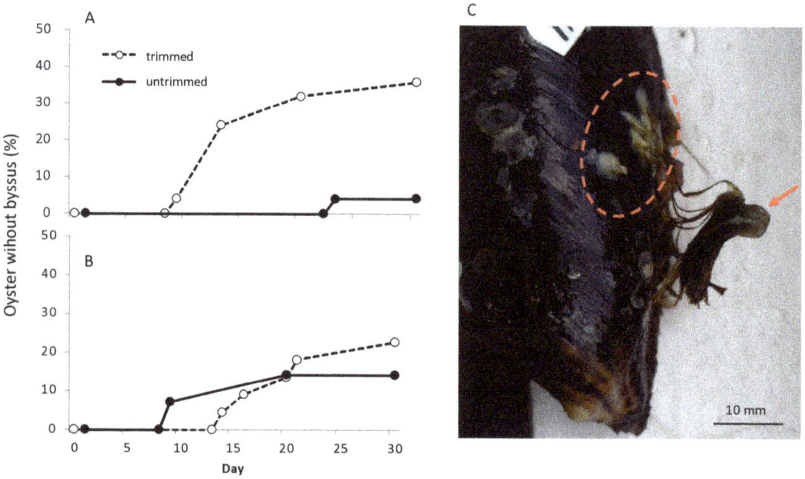

Figure 6. Cumulative percentage of oysters shedding their byssus in the (**A**) small- and (**B**) large-oyster-size groups. Open circles represent oysters that underwent periodic byssal trimming; closed circles represent oysters with untrimmed (intact) byssus. (**C**) Example of a shed byssus from a winged pearl oyster *P. penguin*, following periodic trimming. The arrow indicates the byssal root. The byssus observed in the image belongs to an adjacent oyster and was left attached to the shell surface of the oyster in the image, where individual threads and byssal discs are still visible (dashed oval).

4. Discussion

4.1. Thread Production in Oysters with Trimmed and Untrimmed Byssus

Periodic trimming of byssal threads in *P. penguin* oysters provides valuable insight into byssal production and shedding behavior. Oysters that underwent trimming produced significantly more byssal threads than untrimmed individuals, indicating that trimming stimulates new thread formation. This effect was observed in both small and large oysters (Figure 2), highlighting the consistent impact of trimming across size groups.

The increase in thread counts could be attributed to several factors. Trimming might remove older, less efficient threads, prompting the pearl oyster to produce new ones to maintain its attachment to the substrate. Self-healing of the byssus is not uncommon in bivalve species. *P. nobilis* byssal threads exhibit self-healing behavior through a time-dependent process where hidden length refolds and bonds reform [28]. The *Mytilus* byssal threads demonstrate the capability to disperse as much as 70% of mechanical energy and showcase the potential for self-repair after experiencing pseudoplastic mechanical injury [29,30]. Additionally, the sole action of trimming the byssus in *P. penguin* could potentially stimulate the oyster's physiological processes involved in byssal production, leading to enhanced thread growth. The byssus is a proteinaceous structure that byssate bivalves use for attachment, and its production is influenced by various environmental and physiological factors. Research has shown that environmental stressors, such as heat waves, can impair byssal production in mussels. For instance, exposure to elevated temperatures has been linked to decreased production and quality of byssal threads, affecting their length, diameter, and overall strength [31,32]. Low salinity levels also have been shown to reduce the mussels' byssal production [33]. Research indicates that *Arca boucardi* can form a new byssus within 9–10 h after detachment, producing a structure similar to the original, which highlights the rapid regeneration capability of this species [34]. Similarly, in *M. coruscus*, while high microalgae availability can disrupt byssal secretion, the mussels still exhibit increased byssal production during recovery periods, suggesting resilience in their regenerative processes [35]. Furthermore, *P. penguin* maintains byssal properties under varying temperatures, indicating that environmental factors do not hinder its ability to regenerate [7]. All these findings underscore the regenerative potential of byssus in bivalves, although the efficiency and speed of regeneration may vary based on species and environmental conditions [29,36].

Trimming the byssus in *P. penguin* may induce a stress response that could trigger the pearl oyster's physiological processes to regenerate the byssus. This regeneration process could involve increased metabolic activity and energy allocation towards byssal thread synthesis. Byssal properties are maintained through biochemical processes like polyphenol oxidase (PPO) activity, which is key for byssal formation under stress [7]. As oysters adapt to environmental changes, they increase metabolic activity, focusing energy on byssal thread synthesis. This is supported by transcriptome analysis, showing the role of genes related to protein quality control and cytoskeletal reorganization [7]. The results showed that regardless of the oyster's shell size, periodic trimming significantly increases byssal thread production. This suggests a strong relationship between trimming and enhanced byssal regeneration, indicating that pearl oysters can effectively reattach by producing new byssal threads, regardless of the size of the individual.

Nevertheless, smaller pearl oysters produce significantly more byssal threads than larger ones. Larger oysters may prioritize energy towards maintaining and reinforcing their existing byssal stalks rather than producing new threads. Since heavier oysters already have wider byssal stalks, they might allocate their resources to sustaining the strength and structural integrity of these stalks rather than producing additional threads. In contrast, smaller oysters, which have less massive byssal stalks, may invest more in producing new threads as a way to enhance their attachment strength. Small pearl oysters prioritize byssal development, especially under environmental challenges, while larger ones are less prone to byssal shedding, suggesting limited regeneration as they age [13,23]. The smaller oysters also may be in a different growth phase compared to the larger ones. Small individuals may be in a phase where they naturally produce more byssal threads as part of their growth and development. On the other hand, larger oysters may have passed this phase and instead focus on maintaining their established attachment structures. The larger byssal stalk in large oysters might reach a threshold beyond which the addition of more threads results in diminishing returns, making further increases in byssal diameter less efficient or necessary. This mechanical limitation could explain why larger oysters do not produce as many new threads as small ones, even when their existing threads are trimmed.

The cumulative count of the number of threads produced by each oyster over the observation period reveals a marked difference between the trimmed and untrimmed ones (Figure 3). Pearl oysters subjected to byssal trimming exhibited a consistent and significant increase in byssal thread production over time, with small oysters producing an average of 67.08 $\pm$ 23.31 threads and large oysters producing an average of 43.63 $\pm$ 25.15 threads after up to 10 trimming sessions over 30 days. In contrast, oysters whose byssi were not trimmed displayed less byssal thread production, with small oysters averaging 9.2 $\pm$ 5.2 threads and large oysters averaging 5.28 $\pm$ 6.7 threads. This reinforces the earlier discussed observation that periodic trimming stimulates byssal thread production in pearl oysters. There is evidence that by cutting byssal threads daily, the thread production increased substantially in the zebra mussel *Dreissena polymorpha*. After the byssi were cut, the animals still exhibited the ability to establish reattachment to a substratum [37]. Periodic byssal thread removal also increased thread production in the mussels *M. trossulus* [38] and *M. edulis* [39]. The mussels initially produce about two threads per day during a period of acclimation (7 days). Following this acclimation, they produce between 6 and 12 threads per day. This response indicates that detaching the threads stimulates the mussels to increase their attachment efforts significantly. In the result presented in this investigation, the pearl oysters started to produce more threads after 10 days and continued this trend until the end of the experiment. It is unknown how long they can maintain this trend, particularly because byssal regeneration energy costs usually result in decreased animal growth [38].

4.2. Byssal Diameter in Oysters with Trimmed and Untrimmed Byssus

The examination of the byssal stalk diameter further elucidates the effect of thread trimming on byssal growth (Figure 4). Both small and large oysters exhibited an increase in byssal stalk diameter over the 30-day experimental period. Interestingly, trimming the newly formed byssal threads seemed to stimulate greater byssal diameter compared to leaving its threads intact in both size groups. It seems that trimming the threads at the distal section not only encourages the production of new threads but also 'promotes' the thickening of existing byssal stalk, potentially enhancing the oysters' attachment strength and stability. The merging of the trimmed threads to the byssal stalk seems to be responsible for the increase in the diameter at the proximal section. The correlation between the total number of byssal threads and byssal diameter in adult *P. penguin* oysters confirms this assumption with a significant but modest positive relationship (Figure 5). A potential explanation for this relationship is the merging of newly produced byssal threads with the pre-existing byssal stalk. As more threads are produced, they may fuse and reinforce the existing stalk, leading to an increase in its overall diameter. This merging process could provide structural benefits, enhancing the strength and durability of the byssus, which is crucial for the oyster's ability to anchor securely to substrates. However, the modest correlation (R^2 = 0.19) suggests that while the number of threads contributes to the increase in byssal diameter, other factors are likely involved. Environmental conditions, genetic factors, and the individual health of the oysters could also influence byssal diameter. The relatively low R^2 value implies that the number of threads is only one of several variables affecting the diameter. These findings underscore the complexity of byssal production in *P. penguin*, indicating that the diameter of the byssus is influenced by multiple factors, with the number of threads produced being a significant, but not the sole, determinant.

The exact mechanism of how a single byssal thread merges with the stalk in *P. penguin* is still unknown. The byssus consists of a dense and highly aligned fiber inner core, surrounded by an outer cuticle that contains protein granules embedded in a protein matrix. This structural composition may facilitate the merging of the individual threads into a singular, robust byssus [40]. In various bivalves, the byssal gland in the foot secretes the proteins that form the byssal threads. Multiple threads are simultaneously secreted from the foot. As the threads emerge from the foot, they are guided by grooves and folds in the foot that direct them to converge at a central point [31,41]. The converging threads usually fuse to form a stem-like structure called the byssal stem or peduncle [42]. In *P. nobilis*, two types of threads or filaments are produced: the thick and the thin filaments. The thin filaments are laterally fused in pairs. Two of them form a junction where the other two fit to form the thick filament. A fully developed adult byssus can be formed by 20,000–30,000 filaments [43]. The stem acts as a common attachment point for the individual threads [41]. At the distal end of the stem, the threads splay out and form adhesive plaques that attach to the substrate. The plaques are secreted by the foot and adhere to the surface using specialized proteins. After secretion, the byssal threads and plaques undergo hardening through oxidation and cross-linking of the proteins, increasing the strength and durability of the attachment. Moreover, studies on the biomechanical properties of byssus outside the Mytilidae family have found evidence that the different mechanical properties, such as strength, differ significantly among different groups, although there are some similarities, such as byssal diameter [44].

4.3. Byssal Shedding in Oysters with Trimmed and Untrimmed Byssus

The propensity for byssal shedding provides insight into oyster behavior in response to trimming. Among small oysters with trimmed byssal threads, a higher proportion exhibited byssal shedding compared to those with intact byssus, indicating a potential response to the trimming stimulus (Figure 6A,B). A similar trend was observed in the large-oyster cohort, albeit with a lower shedding rate. Byssal shedding may represent a mechanism for oysters to adjust their attachment in response to environmental conditions or physiological changes induced by trimming. Shedding of byssal stalk or byssal filaments can be attributed to various environmental and biological factors. Excess food availability induces byssal shedding in mussels. Animals fed with high levels of microalgae experienced increased shedding of byssal filaments. High food levels altered metabolic pathways and downregulated genes related to mussel foot proteins, which may explain the increased shedding [35]. Byssal shedding also occurs when animals are stress-free. Control groups in experiments testing the effect of chemical cues from predators or damaged conspecifics of the green-lipped mussel *Perna viridis* exhibited a higher rate of byssal shedding, indicating that stress-free animals were more likely to lose their byssal threads [45]. Changes such as ocean acidification can reduce the mechanical performance of the byssus, resulting in byssal shedding as well [46]. The cause for the byssal shedding seems to be related to the necessity of mobility, allowing reattachment and movement in response to environmental changes [41]. However, the direct influence of extreme environmental factors, such as hyposalinity, or climate change-related factors may cause a reduction in byssal secretion, weakening, or impairment but do not promote detachment [11,33,47,48].

The shedding of the whole byssal stalk, including the byssal root (Figure 6C), in response to periodic trimming of the byssal threads in *P. penguin* may be driven by several factors. One possible reason for shedding the whole byssus is the oysters' ability to recalibrate their attachment strength in response to the trimming of their anchoring threads. Byssal shedding may somehow facilitate the renewal of damaged threads (self-healing), allowing oysters to maintain a robust and functional attachment apparatus over time, as already described in mussels [49,50]. By removing excess threads that are no longer necessary for secure attachment, oysters can also optimize their energy expenditure and adapt to fluctuations in their surroundings. Byssal production is highly energy-intensive, accounting for around 8% of total energy expenditure in *M. edulis*, with significant demands for both carbon and nitrogen [51], accounting for nearly double the rate of oxygen consumption in *P. canaliculus* during byssal attachment [52]. In *M. trossulus*, byssal production significantly impacts energy reserves. In control groups with low byssal thread output, production consumed 2–8% of their energy budget, but in mussels induced to produce threads daily, this increased up to 47% [38]. Despite this demand, energy allocation is prioritized towards byssal thread production over growth [53]. This change in priorities may explain why pearl oysters shed their byssi once the stalk reaches a certain size, as increasing the diameter further becomes less efficient. This idea aligns with previous observations suggesting that byssal growth in these species eventually hits a threshold [13]. However, a clear explanation of how the oyster's byssus is shed after reaching this point is still unclear. A mechanical limitation of space could also trigger a physiological response that explains why oysters promptly exhibit shedding of their byssus after reaching the size threshold. The voluntary and relatively fast release of the entire byssus has recently been described in mussels [14]. Under specific conditions, mussels can shed their byssus cleanly without damaging the stem root, a phenomenon also seen in *P. penguin* (Figure 6C). This clean release is essential for preserving the integrity of the attachment system while enabling mobility when necessary. Serotonin facilitates this release, while dopamine increases the force required for release, highlighting the important role these chemicals play in regulating the organism's ability to detach from surfaces [14].

5. Conclusions

Periodic trimming of byssal threads in *P. penguin* significantly enhances byssal thread production, especially in small-sized oysters, which produce more new threads than their larger counterparts. Trimming stimulates the pearl oyster's regenerative mechanism to maintain substrate attachment, with smaller oysters investing more in thread production, while larger oysters focus on reinforcing their existing byssal stalks. Trimming also increases byssal stalk diameter, likely due to the merging of new threads with existing ones, enhancing attachment strength and stability. Periodic trimming increased both thread production and byssal stalk diameter, suggesting that new threads merge with existing ones to improve structural integrity and attachment strength. Additionally, trimming stimulated byssal shedding, an adaptive response that allows pearl oysters to 'recalibrate' their attachment apparatus. This behavior may optimize energy use, prevent damage to overgrown byssal stalks, and facilitate locomotion in dynamic environments. These findings provide new evidence for the plasticity of attachment strategies in sessile marine invertebrates, with potential applications in aquaculture management and biomaterials research. The byssal shedding behavior in *P. penguin* underscores the dynamic nature of pearl oyster attachment biology and highlights the complex interplay between physiological processes, environmental cues, and adaptive responses. Despite the recent efforts in examining byssal origin, structure, and formation, further research into the molecular mechanisms underlying byssal shedding could provide valuable insights into the regulatory pathways involved and enhance our understanding of pearl oyster ecology and behavior. Such knowledge could inform strategies for improving oyster husbandry practices and resilience in changing marine environments.

Author Contributions: H.E.V.: Investigation, Data Curation, Visualization, Formal Analysis, Writing—Original Draft, and Project Administration. S.W.: Investigation, Conceptualization, Formal Analysis, and Methodology. G.Y.: Investigation and Methodology. L.W.: Investigation and Methodology. P.Y.: Investigation and Methodology. M.D.: Investigation and Methodology. C.Y.: Formal Analysis, Writing—Review and Editing, and Funding Acquisition. X.Z.: Conceptualization, Methodology, Formal Analysis, Writing—Review and Editing, and Funding Acquisition. All authors have read and agreed to the published version of the manuscript.

Funding: This research was funded by the National Key R&D Program of China (Grant No. 2024YFD2401804), the Key Research and Development Project of Hainan Province (Grant No. ZDYF2024XDNY166), Project of Sanya Yazhou Bay Science and Technology City (Grant No. SCKJ-JYRC-2024-50), and the Science and Technology Development Fund Project of the South China Sea Bureau (240101).

Data Availability Statement: The raw data supporting the conclusions of this article will be made available by the authors upon request.

Acknowledgments: We sincerely thank the technicians and management staff at Sanya Tropical Fisheries Research Institute (Lingshui, Hainan Province) for providing the facilities and support necessary to conduct our experiments. Additionally, we acknowledge the use of ChatGPT-4-turbo to enhance the language and readability of this manuscript. The authors have thoroughly reviewed and edited the content and take full responsibility for its accuracy and integrity.

Conflicts of Interest: The authors declare no conflicts of interest.

References

1. Inoue, K.; Onitsuka, Y.; Koito, T. Mussel Biology: From the Byssus to Ecology and Physiology, Including Microplastic Ingestion and Deep-Sea Adaptations. *Fish. Sci.* **2021**, *87*, 761–771. [CrossRef]
2. Simmons, M.; Horbelt, N.; Sverko, T.; Scoppola, E.; Jackson, D.J.; Harrington, M.J. Invasive Mussels Fashion Silk-like Byssus via Mechanical Processing of Massive Horizontally Acquired Coiled Coils. *Proc. Natl. Acad. Sci. USA* **2023**, *120*, e2311901120. [CrossRef] [PubMed]
3. Zhang, Y.; Mao, F.; Xiao, S.; Yu, H.; Xiang, Z.; Xu, F.; Li, J.; Wang, L.; Xiong, Y.; Chen, M. Comparative Genomics Reveals Evolutionary Drivers of Sessile Life and Left-Right Shell Asymmetry in Bivalves. *Genom. Proteom. Bioinform.* **2022**, *20*, 1078–1091. [CrossRef] [PubMed]
4. Wada, K.T.; Tëmkin, I. Taxonomy and Phylogeny. In *The Pearl Oyster*; Elsevier: Amsterdam, The Netherlands, 2008; Volume 2008, pp. 37–75.
5. Gervis, M.H.; Sims, N.A. *The Biology and Culture of Pearl Oysters (Bivalvia pteriidae)*; WorldFish: Penang, Malaysia, 1992; Volume 21, p. 49.
6. Taylor, J.J.; Rose, R.A.; Southgate, P.C. Byssus Production in Six Age Classes of the Silver-Lip Pearl Oyster, *Pinctada maxima* (Jameson). *J. Shellfish Res.* **1997**, *16*, 97–102.
7. Chen, Y.; Han, C.; Chen, H.; Yan, J.; Zhan, X. The Mechanisms Involved in Byssogenesis in *Pteria penguin* under Different Temperatures. *Sci. Total Environ.* **2023**, *905*, 166894. [CrossRef]
8. Li, S.; Liu, C.; Zhan, A.; Xie, L.; Zhang, R. Influencing Mechanism of Ocean Acidification on Byssus Performance in the Pearl Oyster *Pinctada fucata*. *Environ. Sci. Technol.* **2017**, *51*, 7696–7706. [CrossRef]
9. Welladsen, H.M.; Heimann, K.; Southgate, P.C. The Effects of Exposure to Near-Future Levels of Ocean Acidification on Activity and Byssus Production of the Akoya Pearl Oyster, *Pinctada Fucata*. *J. Shellfish Res.* **2011**, *30*, 85–88. [CrossRef]
10. O'Connor, W.A.; Lawler, N.F. Salinity and Temperature Tolerance of Embryos and Juveniles of the Pearl Oyster, *Pinctada imbricata* Röding. *Aquaculture* **2004**, *229*, 493–506. [CrossRef]
11. Vasquez, H.E.; Xing, Z.; Zhan, X.; Gu, Z.; Wang, A. Byssal Re-attachment Behavior in the Winged Pearl Oyster *Pteria penguin* in Response to Low Salinity Levels. *J. World Aquac. Soc.* **2021**, *52*, 457–465. [CrossRef]

12. Kishore, P.; Hunter, J.; Zeng, C.; Southgate, P.C. The Effects of Different Culture Apparatuses and Current Velocities on Byssus Production by the Black-Lip Pearl Oyster, *Pinctada margaritifera*. *Aquaculture* **2014**, *434*, 74–77. [CrossRef]
13. Vasquez, H.E.; Zheng, X.; Zhan, X.; Gu, Z.; Wang, A. Byssus Growth in Winged Pearl Oyster *Pteria penguin* (Röding, 1798). *J. Shellfish Res.* **2018**, *37*, 515–519. [CrossRef]
14. Sivasundarampillai, J.; Youssef, L.; Priemel, T.; Mikulin, S.; Eren, E.D.; Zaslansky, P.; Jehle, F.; Harrington, M.J. A Strong Quick-Release Biointerface in Mussels Mediated by Serotonergic Cilia-Based Adhesion. *Science* **2023**, *382*, 829–834. [CrossRef] [PubMed]
15. Dharmaraj, S.; Kandasamy, D.; Alagarswami, K. Some Aspects of Physiology of Pearl Oyster. *CMFRI Bull.-Pearl Cult.* **1987**, *39*, 21–28.
16. Giraldes, B.W.; Leitão, A.; Smyth, D. The Benthic Sea-Silk-Thread Displacement of a Sessile Bivalve, *Pinctada imbricata radiata* (Leach, 1819) in the Arabian-Persian Gulf. *PLoS ONE* **2019**, *14*, e0215865. [CrossRef]
17. Nicastro, K.; Zardi, G.; McQuaid, C. Movement Behaviour and Mortality in Invasive and Indigenous Mussels: Resilience and Resistance Strategies at Different Spatial Scales. *Mar. Ecol. Prog. Ser.* **2008**, *372*, 119–126. [CrossRef]
18. Liu, J.; Li, Q.; Kong, L.; Yu, H.; Zheng, X. Identifying the True Oysters (Bivalvia: Ostreidae) with Mitochondrial Phylogeny and Distance-based DNA Barcoding. *Mol. Ecol. Resour.* **2011**, *11*, 820–830. [CrossRef] [PubMed]
19. Johnston, W.; Gordon, S.E.; Wingfield, M.; Halafihi, T.; Southgate, P.C. Influence of Production Method on the Profitability of Mabé Pearl Farming Using Traditional and Research-Informed Nucleus Implanting Practices with the Winged Pearl Oyster, *Pteria penguin*. *Aquaculture* **2022**, *546*, 737280. [CrossRef]
20. Nur, I.; Mushaffa, W.O.; Hamzah, M. Effect of Number of Nuclei and Nucleus Position on Shell Growth and Mabé Pearl Coating in *Pteria penguin* Cultured in Coastal Waters of Southeast Sulawesi, Indonesia. *J. Shellfish Res.* **2020**, *39*, 345. [CrossRef]
21. Kishore, P.; Southgate, P.C. A Detailed Description of Pearl-Sac Development in the Black-Lip Pearl Oyster, *Pinctada margaritifera* (Linnaeus 1758). *Aquac. Res.* **2016**, *47*, 2215–2226. [CrossRef]
22. Tëmkin, I. Morphological Perspective on the Classification and Evolution of Recent Pterioidea (Mollusca: Bivalvia). *Zool. J. Linn. Soc.* **2006**, *148*, 253–312. [CrossRef]
23. Vasquez, H.E.; Zheng, X.; Gu, Z.; Wang, A. Relationships between Shell Morphological Traits and the Byssus Dimensions in the Winged Pearl Oyster *Pteria penguin* (Röding, 1798) Cultivated in Sanya, Hainan Island, China. *J. Shellfish Res.* **2017**, *36*, 669–676. [CrossRef]
24. Patil, I. Visualizations with Statistical Details: The "ggstatsplot" Approach. *J. Open Source Softw.* **2021**, *6*, 3167. [CrossRef]
25. Wickham, H. *Ggplot2: Elegant Graphics for Data Analysis*; Springer New York: New York, NY, USA, 2016; ISBN 978-0-387-98140-6.
26. Ogle, D.; Doll, J.; Wheeler, A.; Dinno, A. *Package: FSA*, version 0.9.6; FSA: Simple Fisheries Stock Assessment Methods; fishR Core Team: Vienna, Austria, 2025.
27. R Core Team. *R: A Language and Environment for Statistical Computing*. R Foundation for Statistical Computing: Vienna, Austria, 2021. Available online: https://www.R-Project.org/ (accessed on 1 December 2024).
28. Pasche, D.; Horbelt, N.; Marin, F.; Motreuil, S.; Fratzl, P.; Harrington, M.J. Self-Healing Silk from the Sea: Role of Helical Hierarchical Structure in *Pinna nobilis* Byssus Mechanics. *Soft Matter* **2019**, *15*, 9654–9664. [CrossRef]
29. Harrington, M.J.; Waite, J.H. Holdfast Heroics: Comparing the Molecular and Mechanical Properties of *Mytilus californianus* Byssal Threads. *J. Exp. Biol.* **2007**, *210*, 4307–4318. [CrossRef] [PubMed]
30. Reinecke, A.; Bertinetti, L.; Fratzl, P.; Harrington, M.J. Cooperative Behavior of a Sacrificial Bond Network and Elastic Framework in Providing Self-Healing Capacity in Mussel Byssal Threads. *J. Struct. Biol.* **2016**, *196*, 329–339. [CrossRef] [PubMed]
31. Li, S.; Xia, Z.; Chen, Y.; Gao, Y.; Zhan, A. Byssus Structure and Protein Composition in the Highly Invasive Fouling Mussel *Limnoperna fortunei*. *Front. Physiol.* **2018**, *9*, 418. [CrossRef]
32. Xu, X.; Yang, K.; Liu, Y.; Deng, Y.; Zhao, L. Heatwaves Hinder Mussel Invasion by Weakening Byssus Production. *Front. Mar. Sci.* **2023**, *10*, 1239801. [CrossRef]
33. Wang, W.-Y.; Ni, J.-Y.; Huang, S.-H.; Cui, Q.-W.; Wang, Y.-Q.; Gu, Z.-Q.; Li, Y.-F. Hyposalinity Stress Reduces Mussel Byssus Secretion but Does Not Cause Detachment. *Sci. Total Environ.* **2024**, *930*, 172561. [CrossRef]
34. Selin, N.I. The Byssus Morphology and Strength of Attachment to Substrate in the Bivalve Mollusk *Arca boucardi* Jousseaume, 1894 (Arcidae). *Russ. J. Mar. Biol.* **2021**, *47*, 403–406. [CrossRef]
35. Ni, J.-Y.; Zhou, Y.; Wang, Y.-Q.; Huang, S.-H.; Cui, Q.-W.; Wang, W.-Y.; Yang, X.-Y.; Power, D.M.; Li, Y.-F. Examination of the Effects of Excess Microalgae Availability on the Disruption of Mussel Byssus Secretion. *Aquaculture* **2024**, *590*, 741106. [CrossRef]
36. Padin, X.A.; Babarro, J.M.F.; Silva, E.; Longa Portabales, M.Á.; Calvo, S.; Nolasco, R. Variability in Strength of Byssus Attachment and Index Condition of Subtidal Mussels during the Maximum Growth Stage. *Aquac. Res.* **2021**, *52*, 3485–3497. [CrossRef]
37. Rajagopal, S.; Velde, G.; Jenner, H.A.; Gaag, M.; Kempers, A.J. Effects of Temperature, Salinity and Agitation on Byssus Thread Formation of Zebra Mussel *Dreissena polymorpha*. *Neth. J. Aquat. Ecol.* **1996**, *30*, 187–195. [CrossRef]

38. Roberts, E.A.; Newcomb, L.A.; McCartha, M.M.; Harrington, K.J.; LaFramboise, S.A.; Carrington, E.; Sebens, K.P. Resource Allocation to a Structural Biomaterial: Induced Production of Byssal Threads Decreases Growth of a Marine Mussel. *Funct. Ecol.* **2021**, *35*, 1222–1239. [CrossRef]
39. Young, G.A. Byssus-Thread Formation by the Mussel *Mytilus edulis*: Effects of Environmental Factors. *Mar. Ecol. Prog. Ser.* **1985**, *24*, 261–271. [CrossRef]
40. Chen, Y.; Chen, H.; Han, C.; Ou, H.; Zhan, X. The Structure and Proteomic Analysis of Byssus in *Pteria penguin*: Insights into Byssus Evolution and Formation. *J. Proteom.* **2024**, *307*, 105267. [CrossRef]
41. McCartney, M.A. Structure, Function and Parallel Evolution of the Bivalve Byssus, with Insights from Proteomes and the Zebra Mussel Genome. *Philos. Trans. R. Soc. B Biol. Sci.* **2021**, *376*, 20200155. [CrossRef]
42. Waite, J.H.; Harrington, M.J. Following the Thread: *Mytilus* Mussel Byssus as an Inspired Multi-Functional Biomaterial. *Can. J. Chem.* **2022**, *100*, 197–211. [CrossRef]
43. Basso, L.; Vázquez-Luis, M.; García-March, J.R.; Deudero, S.; Alvarez, E.; Vicente, N.; Duarte, C.M.; Hendriks, I.E. The Pen Shell, *Pinna nobilis*. In *Advances in Marine Biology*; Elsevier: Amsterdam, The Netherlands, 2015; Volume 71, pp. 109–160. ISBN 978-0-12-803305-0.
44. Pearce, T.; LaBarbera, M. Biomechanics of Byssal Threads Outside the Mytilidae: *Atrina rigida* and *Ctenoides mitis*. *J. Exp. Biol.* **2009**, *212*, 1449–1454. [CrossRef]
45. Cheung, S.G.; Tong, P.Y.; Yip, K.M.; Shin, P.K.S. Chemical Cues from Predators and Damaged Conspecifics Affect Byssus Production in the Green-Lipped Mussel *Perna viridis*. *Mar. Freshw. Behav. Physiol.* **2004**, *37*, 127–135. [CrossRef]
46. O'Donnell, M.J.; George, M.N.; Carrington, E. Mussel Byssus Attachment Weakened by Ocean Acidification. *Nat. Clim. Change* **2013**, *3*, 587–590. [CrossRef]
47. Li, S. Dietary Exposure to $nTiO_2$ Reduces Byssus Performance of Mussels under Ocean Warming. *Sci. Total Environ.* **2023**, *881*, 163499. [CrossRef] [PubMed]
48. Xiong, X.; Quan, F.; Zhang, Y.; Cao, Y.; Zhang, K.; Li, W.; Luo, H.; Jia, Y.; Liao, Z.; Liu, X. Effect of Low-Salt on the Survival of Mussel *Mytilus coruscus* and Its Molecular Responses to Chronic Prolonged Low-Salt Stress. *Aquaculture* **2024**, *585*, 740689. [CrossRef]
49. Harrington, M.J.; Gupta, H.S.; Fratzl, P.; Waite, J.H. Collagen Insulated from Tensile Damage by Domains That Unfold Reversibly: In Situ X-Ray Investigation of Mechanical Yield and Damage Repair in the Mussel Byssus. *J. Struct. Biol.* **2009**, *167*, 47–54. [CrossRef]
50. Krauss, S.; Metzger, T.H.; Fratzl, P.; Harrington, M.J. Self-Repair of a Biological Fiber Guided by an Ordered Elastic Framework. *Biomacromolecules* **2013**, *14*, 1520–1528. [CrossRef] [PubMed]
51. Bayne, B.L. Evolution. In *Developments in Aquaculture and Fisheries Science*; Elsevier: Amsterdam, The Netherlands, 2017; Volume 41, pp. 47–87.
52. Lurman, G.J.; Hilton, Z.; Ragg, N.L.C. Energetics of Byssus Attachment and Feeding in the Green-Lipped Mussel *Perna canaliculus*. *Biol. Bull.* **2013**, *224*, 79–88. [CrossRef]
53. Roberts, E.A.; Carrington, E. Byssal Thread Attachment and Growth Are Not Correlated across Gradients of Temperature and Food Availability for Two Congeneric Mussel Species. *Mar. Ecol. Prog. Ser.* **2023**, *704*, 35–54. [CrossRef]

Journal of
*Marine Science
and Engineering*

MDPI

Article

Semi-Supervised Underwater Image Enhancement Method Using Multimodal Features and Dynamic Quality Repository

Mu Ding [1,2], Gen Li [1], Yu Hu [1], Hangfei Liu [1], Qingsong Hu [2] and Xiaohua Huang [1,*]

1 South China Sea Fisheries Research Institute, Chinese Academy of Fishery Sciences/Key Laboratory for Sustainable Utilization of Open-Sea Fishery, Ministry of Agriculture and Rural Affairs, Guangzhou 510300, China; huyu@scsfri.ac.cn (Y.H.)
2 College of Engineering Science and Technology, Shanghai Ocean University, Shanghai 201306, China; qshu@shou.edu.cn
* Correspondence: huangxhua@scsfri.ac.cn

Abstract: Obtaining clear underwater images is crucial for smart aquaculture, so it is necessary to repair degraded underwater images. Although underwater image restoration techniques have achieved remarkable results in recent years, the scarcity of labeled data poses a significant challenge to continued advancement. It is well known that semi-supervised learning can make use of unlabeled data. In this study, we proposed a semi-supervised underwater image enhancement method, MCR-UIE, which utilized multimodal contrastive learning and a dynamic quality reliability repository to leverage the unlabeled data during training. This approach used multimodal feature contrast regularization to prevent the overfitting of incorrect labels, and secondly, introduced a dynamic quality reliability repository to update the output as pseudo ground truth. The robustness and generalization of the model in pseudo-label generation and unlabeled data learning were improved. Extensive experiments conducted on the UIEB and LSUI datasets demonstrated that the proposed method consistently outperformed existing traditional and deep learning-based approaches in both quantitative and qualitative evaluations. Furthermore, its successful application to images captured from deep-sea cage aquaculture environments validated its practical value. These results indicated that MCR-UIE held strong potential for real-world deployment in intelligent monitoring and visual perception tasks in complex underwater scenarios.

Keywords: aquaculture; underwater image enhancement; multimodal contrastive learning; dynamic quality reliability repository

1. Introduction

Driven by the expansion of marine resource exploitation and the advancement of oceanographic research, underwater images have progressively gained importance in fields such as marine surveillance, underwater archeology, and aquaculture. Nevertheless, the unique optical properties of underwater environments, such as light absorption, scattering, and refraction, often lead to severe image degradation, including color distortion, low contrast, and detail blurring. These degradations significantly impair both the visual perception and the effectiveness of subsequent image analysis. Underwater image enhancement seeks to address these challenges by improving visual quality, thereby facilitating human interpretation and automated processing [1,2]. Through enhancement processing, the sharpness, contrast, and color fidelity of images can be increased, which plays a vital role in tasks encompassing underwater target detection, recognition, and tracking.

https://doi.org/10.3390/jmse13061195

Researchers have explored various enhancement techniques, each with their own set of advantages and limitations. Currently, underwater image enhancement approaches are typically classified into three main categories: physical model-based approaches [3,4], non-physical model-based approaches [5,6], and deep learning-based approaches [7–9].

Physical model-based: These methods describe the degradation process of underwater images through mathematical deduction, which mainly includes establishing an underwater optical imaging model, estimating the parameters of the model, and restoring underwater images through reverse deduction. The physical parameters in the model are obtained through two different methods: extraction by polarization imaging equipment and derivation by prior knowledge. However, since the parameter estimation process has difficulty fully considering different water environments and shooting conditions, the algorithm is not universal enough.

Non-physical model-based: These approaches based on non-physical models directly adjust the pixel values of the entire image by constructing functions to adjust the color and contrast of the image, thereby enhancing human subjective visual perception. The enhancement approach based on the non-physical model is relatively simple, with low computational complexity, and is easy to implement and apply. However, since it does not take into account the physical process and optical properties, it has poor adaptability to changes in underwater environments and different scenes. It is easy for noise and color deviation to be introduced during processing, and there are problems such as vignetting and oversaturation which affect the quality and clarity of the image.

Underwater image enhancement methods grounded in physical models need to be evaluated and optimized according to specific circumstances in practical applications, which also leads to the insufficient robustness of the methods. Underwater image enhancement approaches grounded in non-physical models also have shortcomings such as poor robustness, limited processing effects, and the presence of artifacts and noise, and can be considered as auxiliary methods. With the continuous development of image processing technology, deep learning-based approaches have gradually addressed some of the shortcomings of traditional methods.

Despite their respective advantages, both physical and non-physical methods suffer from generalization issues when applied to real-world underwater environments. Recent advances in computer vision have spurred the development of deep learning-based methods [10–12], which leverage powerful neural architectures including convolutional neural network (CNN), generative adversarial network (GAN), and autoencoders to learn the mappings from degraded to enhanced images. These approaches have shown promising results, especially when trained on large annotated datasets. However, their performance is highly dependent on the availability of labeled data, which is costly and labor-intensive to obtain in underwater domains.

In contrast, while unlabeled underwater images are relatively easy to collect, the primary challenge lies in how to utilize them effectively to train robust and generalizable models [13]. Existing supervised underwater image enhancement (UIE) methods heavily depend on paired data or high-quality reference images, which are difficult to obtain in real-world underwater environments. Moreover, conventional semi-supervised learning methods, although promising, often fall short in underwater applications due to their reliance on static pseudo-labels and heuristic confidence mechanisms that cannot adapt to sample-wise quality variations or unknown degradations.

To address these limitations, we introduce a semi-supervised learning framework specifically tailored for underwater image enhancement, aiming to improve the model's generalization to diverse and unseen real-world underwater scenarios. Our method is built upon the mean teacher paradigm [14,15], which leverages an exponential moving average (EMA) of the student model to form the teacher network. The teacher provides pseudo-labels for the unlabeled data, and a consistency loss is used to guide the student's training, enabling the model to benefit from both labeled and unlabeled samples.

However, directly applying the mean teacher method to underwater image enhancement poses several critical challenges. (1) The teacher model, especially in the early training stages, is not guaranteed to outperform the student, leading to unreliable pseudo-labels that can misguide the student and hinder convergence. (2) The use of a conventional pixel-wise consistency loss (typically L1 loss) can be overly strict, causing the model to overfit noisy pseudo-labels and suffer from confirmation bias. These issues highlight a scientific gap: understanding how to integrate pseudo-label selection with image quality awareness in a dynamically evolving underwater learning scenario.

To this end, we propose a novel dynamic quality reliability repository (DQR), which continuously tracks and stores high-quality outputs from the teacher model using an NR-IQA metric (MUSIQ). This allows the student to be guided by only the most reliable pseudo-labels, effectively filtering out noisy supervision and stabilizing the semi-supervised process. Furthermore, to alleviate overfitting and enforce a more flexible learning objective, we introduce a multimodal contrastive loss that leverages complementary modality cues, such as VGG features, edge information, color distributions, and local texture regions, to provide gradient-level supervision. This auxiliary contrastive regularization improves representation robustness and is especially beneficial when working with unlabeled, degraded underwater images.

Taken together, our proposed approach directly addresses the shortcomings of previous semi-supervised UIE methods by integrating dynamic reliability assessment and multimodal regularization. It offers a principled strategy to fully exploit large-scale unlabeled underwater data, filling an important methodological gap and improving model generalization across a wide range of underwater environments.

The primary contributions of this work are summarized as follows: (1) We proposed MCR-UIE, a semi-supervised underwater image enhancement framework that leveraged multimodal loss and a dynamic quality reliability repository to effectively utilize unlabeled data, thereby improving the generalization capability of the trained model on real-world images. (2) To ensure the reliability of the pseudo-labels, we constructed a dynamic quality reliability repository that archived the best outputs produced by the teacher model. (3) We adopted multimodal contrastive loss as a regularization technique to alleviate confirmation bias during training. (4) Extensive experimental results demonstrated the effectiveness and robustness of the proposed approach.

The remainder of this manuscript is organized as follows: Section 2 reviews the related work. In Section 3, we introduce the proposed semi-supervised underwater image enhancement method, which incorporates multimodal contrastive loss and the dynamic quality reliability repository. Section 4 presents the enhanced experimental and analytical results. Finally, the conclusions are summarized in Section 5.

2. Related Works

2.1. Underwater Image Enhancement Methods

Traditional underwater image enhancement approaches are generally categorized into physical and non-physical model-based methods. Physical model-based approaches [3,4] aim to describe the image degradation process by estimating unknown parameters in underwater imaging models. These parameters typically include the transmission map and ambient light, which are derived using handcrafted priors and assumptions based on optical principles. In contrast, non-physical model-based approaches directly improve image quality by adjusting pixel intensities or contrast through algorithmic design. Typical techniques include CLAHE [6], Retinex [5], Fusion [16], and MMLE [17]. Although these traditional methods have achieved reasonable performance in relatively simple underwater scenes, they often struggle to handle complex and dynamic real-world environments. Their limitations become evident when facing the varying lighting conditions, turbidity levels, and color distortions that are common in practical underwater applications.

Early deep learning-based underwater image enhancement approaches [18,19] commonly relied on physical imaging models. These methods typically trained neural networks to estimate parameters such as transmission maps and ambient light. However, the challenge of accurately estimating these parameters often led to suboptimal restoration performance, particularly in complex underwater environments.

To overcome these limitations, recent research has shifted towards purely data-driven approaches that dispense with explicit imaging models. These approaches aim to learn a direct mapping from degraded to enhanced images using supervised learning on paired datasets. For example, some frameworks employ feature fusion strategies that integrate outputs from multiple traditional enhancement methods to guide the restoration process [20]. Others incorporate prior-inspired modules, such as spatial encoders and transmission-guided decoders, to refine structural and color representations [21]. Additionally, GAN-based architectures have also been introduced to achieve efficient, real-time image enhancement [22].

2.2. Semi-Supervised Approaches

In recent times, semi-supervised learning has become an effective strategy in computer vision by enabling the joint utilization of labeled and unlabeled data. Several representative approaches have been proposed, including mean teacher [14], virtual adversarial learning [23], and Fixmatch [24]. Among these approaches, the mean teacher method, grounded in consistency regularization, has shown remarkable effectiveness in image classification tasks. Its effectiveness has subsequently inspired its application in other areas such as semantic segmentation [25] and image restoration [26].

Despite the increasing popularity of semi-supervised learning in various vision-related tasks, its potential in underwater image restoration remains largely unexplored. A preliminary study [27] attempted to apply a semi-supervised strategy by jointly optimizing supervised and unsupervised losses within a single network. Building upon this idea, our work proposes a more systematic framework that incorporates several key components tailored for underwater scenarios. Specifically, we adopt the mean teacher mechanism and further enhance it with a dynamic quality reliability repository to filter pseudo-labels, as well as a multimodal contrastive loss that promotes better feature representation and mitigates confirmation bias.

2.3. Contrastive Learning

Contrastive learning represents a powerful paradigm in self-supervised representation learning [28]. It facilitates the acquisition of meaningful visual features by enforcing similarity between semantically related samples while pushing apart dissimilar ones. In the domain of image restoration, previous studies primarily focus on constructing contrastive pairs and designing appropriate feature projections. For instance, some approaches [29,30] consider clean images as positive samples and degraded ones as negatives, projecting them into a learned embedding space using networks such as VGG [31]. However, these implementations typically rely on paired ground truth and apply the contrastive loss in a supervised manner, limiting their applicability to unlabeled data.

To date, contrastive learning has seen limited use in underwater image restoration. A prior work [32] incorporated contrastive loss as a regularization term to improve performance within a supervised learning framework, but it still depended on labeled data.

In contrast, this study presents a systematic approach to utilizing unlabeled data through multimodal contrastive learning. By designing contrastive objectives that leverage information from multiple modalities, our method enables the network to learn more robust features without requiring ground truth, thereby enhancing its generalization to complex real-world underwater scenes.

3. Methods

3.1. The Network Structure of MCR-UIE

Semi-supervised learning is intended to leverage both labeled and unlabeled data to improve model generalization and learning efficiency. In the context of underwater image restoration, we formally define the problem as follows: Let the labeled dataset be denoted as $D_L = \left\{\left(x_i^l, y_i^l\right)\middle| x_i^l \in I_l^{LQ}, y_i^l \in I_l^{HQ}\right\}_{i=1}^N$, where x_i^l and y_i^l represent the degraded underwater image and its corresponding clean ground truth, respectively, sampled from the low-quality set I_l^{LQ} and the high-quality set I_l^{HQ}. Similarly, the unlabeled dataset is defined as $D_U = \left\{x_i^u \middle| x_i^u \in I_u^{LQ}\right\}_{i=1}^M$, where each x_i^u is an underwater image from the unlabeled degraded set I_u^{LQ}. It is important to note that the labeled and unlabeled images are disjointed, i.e., $D_L \cap D_U = \emptyset$. The overall objective is to learn a restoration mapping function over the combined dataset $D = D_L \cup D_U$, such that any degraded underwater image x can be effectively transformed into its clean version y.

Our semi-supervised learning framework adopts the standard architecture commonly used in semi-supervised settings [14,24], as presented in Figure 1. Specifically, the proposed MCR-UIR consists of two networks with identical architectures, referred to as the teacher and student networks. The key distinction between them lies in their parameter update strategy: the student network is trained via gradient descent, while the teacher network is updated using the exponential moving average of the student's weights during training.

The teacher network's parameters, denoted as θ_t, are refined using the EMA of the student network's parameters θ_s, following the update rule:

$$\theta_t = \lambda\theta_t + (1 - \lambda)\theta_s \tag{1}$$

where $\lambda \in (0,\ 1)$ is a momentum coefficient that controls the update speed. This strategy enables the teacher model to accumulate knowledge from the student network over time, effectively aggregating its parameters after each training step. As highlighted in [33], such temporal weight averaging not only helps to stabilize the training process but also leads to better generalization performance compared to standard gradient descent.

The student network's parameters are refined via gradient descent. Typically, the optimization objective of the student model is defined by minimizing the following loss function:

$$L_{total} = L_{sup} + \lambda L_{unsup} \tag{2}$$

where $L_{sup} = \sum_{i=0}^N \left|f_{\theta_s}\left(x_i^l\right) - y_i^l\right|$ denotes the supervised loss, and $L_{unsup} = \sum_{i=0}^M \left|f_{\theta_s}(\phi_s(x_i^u)) - f_{\theta_t}(\phi_t(x_i^u))\right|$ represents the unsupervised student–teacher consistency loss. Here, $|\cdot|$ denotes the L1 distance, and ϕ_s and ϕ_t refer to the data augmentation functions applied to the student and teacher inputs, respectively.

In principle, as the teacher network generally yields superior performance, the unsupervised loss L_{unsup} provides effective guidance for training the student model on unlabeled samples. Accordingly, the teacher's output $\hat{y}_i^u = f_{\theta_t}(\phi_t(x_i^u))$ is referred to as a pseudo-label. However, it is important to note that the teacher's predictions are not always more accurate than those of the student. Inaccurate pseudo-labels may introduce noise and negatively impact the learning process of the student network.

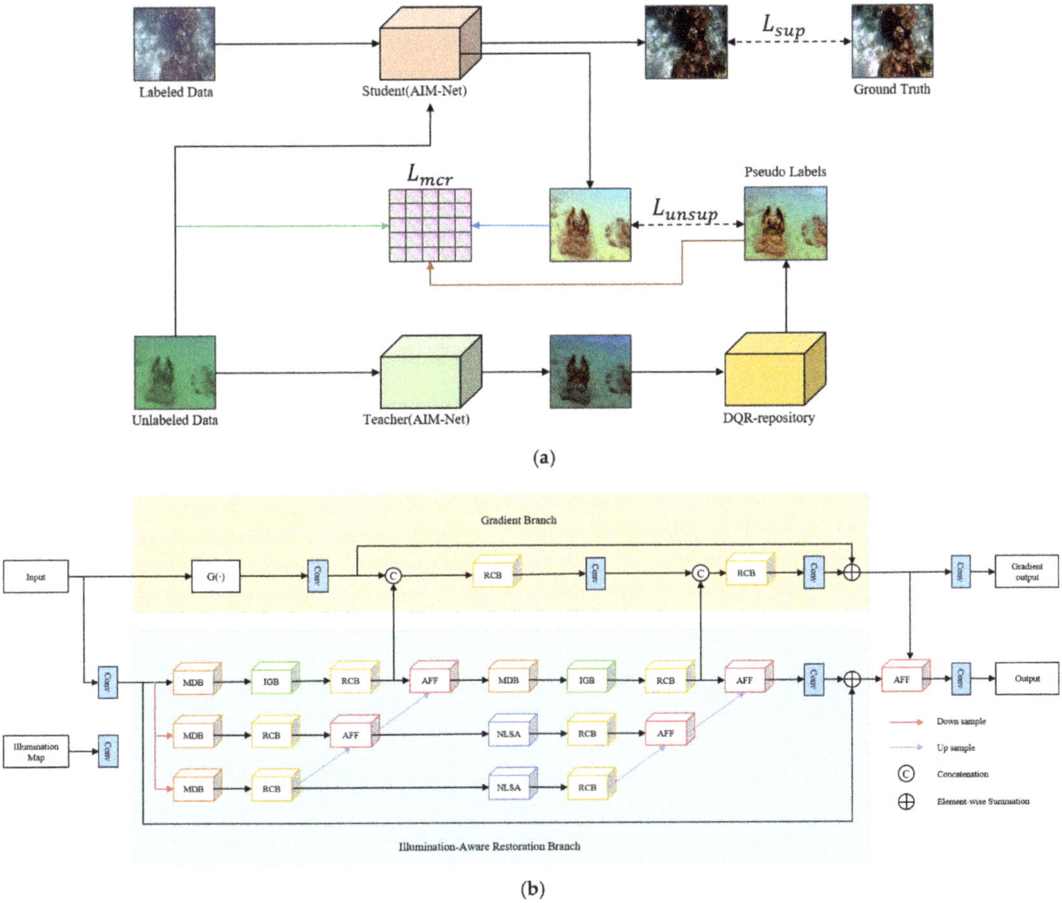

Figure 1. Illustration of our MCR-UIE framework. (**a**) The network structure of the proposed MCR-UIE; (**b**) the Asymmetric Illumination-Aware Multi-Scale Network.

3.2. Dynamic Quality and Reliability Repository

To tackle the aforementioned problem, we utilize the most confident outputs from the teacher network as pseudo-labels. Similar approaches have been employed in image classification and semantic segmentation tasks [25]; output reliability is typically assessed based on prediction entropy or confidence scores. However, directly extending these approaches to image restoration tasks is non-trivial due to several unique challenges. Specifically, as a regression problem, underwater image restoration requires the accurate recovery of fine textures and the effective removal of color casts, which cannot be reliably evaluated using classification-based confidence metrics.

To mitigate the issue of unreliable pseudo-labels, we design a reliable repository that dynamically stores the most trustworthy outputs generated by the teacher network during training. Initially, the repository B_U is empty. At each iteration, we evaluate the current output of the teacher model by comparing it with both the corresponding output of the student model and the existing pseudo-label in the repository. If the teacher's current prediction exhibits better quality, it replaces the previous one in B_U. As training proceeds, the repository gradually accumulates a set of reliable pseudo-labels, denoted as $B_U = \left\{y_i^b\right\}_{i=1}^{M}$. In this way, we construct the updated pseudo-labeled dataset $D' = D_U \cup B_U = \left\{ \left(x_i^u, y_i^b \right) \right\}_{i=1}^{M}$, where each unlabeled image is associated with its most reliable pseudo-label. This mechanism ensures that the unsupervised consistency loss L_{unsup} is computed using high-quality targets, thereby reducing the adverse effects of noisy supervision. The revised loss function can be formulated as follows:

$$L'_{unsup} = \sum_{i=0}^{M} \left| f_{\theta_s}(\phi_s(x_i^u)) - y_i^b \right| \tag{3}$$

Intuitively, one might consider using non-reference image quality assessment (NR-IQA) metrics. However, as highlighted in [3,20], widely used metrics such as UCIQE [34] and UIQM [35] do not reliably capture the quality of restored underwater images. Consequently, relying on these metrics to construct our dynamic quality and reliability repository could lead to suboptimal results. To address this issue, we perform an empirical evaluation of multiple NR-IQA metrics to identify the most suitable one for assessing the quality of underwater images. We observe that the deep learning-based MUSIQ [36] metric best aligns with the monotonicity law. To justify the use of MUSIQ as the reliability criterion in our dynamic quality repository, we conduct a comparative study on seven commonly used NR-IQA methods over the EUVP benchmark, which covers a wide range of underwater scenarios. As shown in Figure 2, the evaluation highlights that deep learning-based metrics, particularly MUSIQ, exhibit better monotonicity and alignment with visual quality as compared to traditional handcrafted metrics such as BRISQUE and NIQE. MUSIQ consistently provides more stable and perceptually meaningful scores across varying underwater degradations. Based on this observation, we adopt MUSIQ to estimate the reliability of the network outputs, guiding the update of pseudo-labels in our dynamic quality reliability repository.

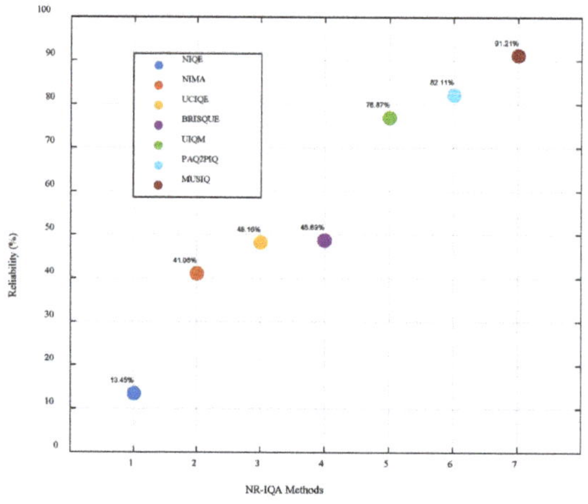

Figure 2. Comparison of seven NR-IQA metrics on the EUVP benchmark. MUSIQ shows the best consistency with visual quality perception, validating its suitability for underwater pseudo-label selection.

The construction steps of the dynamic quality reliability repository are detailed in Algorithm 1, with corresponding explanations provided for each step. Obtain the predictions of teachers and students: We calculate the prediction results of the teacher and student models for unlabeled samples. The predictions of the teacher model are used to generate candidate pseudo-labels, and the predictions of the student model are used for comparative judgment. Segment local areas: We divide the prediction results into multiple small blocks to evaluate the quality of the image more carefully, so as to avoid the situation where the overall score is affected by the poor local quality of the image. Calculate local quality scores and entropy metrics: We calculate the NR-IQA score of each local area separately, and use the entropy metric to measure the uncertainty of the prediction, and then weight the local scores into a global score. The smaller the entropy, the more confident the model prediction is, so a negative weight is given to the entropy metric. Update the reliable sample repository: If the prediction quality score of the teacher model is significantly higher than that of the student model and exceeds the existing pseudo-labels in the repository, it means that the pseudo-label is more reliable and can be replaced in the reliable sample repository. We adopt an online update mechanism for the dynamic quality reliability repository, where the repository is dynamically refreshed during each training iteration. For every unlabeled input image, we generate enhancement results from both the teacher and student branches and assess their quality using a composite reliability score that integrates the MUSIQ score and entropy-based confidence. A new prediction from the teacher branch is allowed to replace an existing sample in the repository only when it achieves a higher quality score than both the corresponding student output and the current repository entry. This selective replacement ensures that only more reliable and higher quality pseudo-labels are preserved, allowing the DQR to evolve towards greater consistency and trustworthiness over time.

To ensure stable convergence in this feedback-based design, we apply a progressive warm-up strategy to the consistency regularization coefficient ρ, which gradually increases its influence throughout training. In the early stages, when pseudo-labels may still be noisy, this scheduling helps prevent unstable updates. As the network matures, fewer replacements occur, and the repository contents stabilize, thereby promoting convergence. This interplay between dynamic updating and progressive regularization contributes to the robustness and effectiveness of our semi-supervised enhancement framework.

Algorithm 1: Update of dynamic quality reliability repository

Require: NR-IQA method $\Psi(\cdot)$, Entropy metric $H(\cdot)$, Local region split function $split(\cdot)$;
Initialize: $B_U = \varnothing$;
Sample a batch of unlabeled images $\left\{x_i^u\right\}_{i=1}^b$ from D_U;
for each x_i^u **do**
  Get teacher's prediction: $\hat{y}_i^u = f_{\theta_t}(\phi_t(x_i^u))$;
  Get student's prediction: $\tilde{y}_i^u = f_{\theta_s}(\phi_s(x_i^u))$;
  Compute enhanced quality scores for $\hat{y}_i^u$, $\tilde{y}_i^u$, and $y_i^b \in B_U$;
  Split each prediction into local regions: $\{r_k(\hat{y}_i^u)\}$, $\left\{r_k\left(\tilde{y}_i^u\right)\right\}$, $\left\{r_k\left(y_i^b\right)\right\}$ using $split(\cdot)$;
  Compute NR-IQA scores of each region for teacher prediction:
  $z_t^k = \Psi(r_k(\hat{y}_i^u))$;
  Compute NR-IQA scores of each region for student prediction:
  $z_s^k = \Psi\left(r_k\left(\tilde{y}_i^u\right)\right)$;
  Compute NR-IQA scores of each region for existing reliable bank sample:
  $z_b^k = \Psi\left(r_k\left(y_i^b\right)\right)$.
  Aggregate regional scores with weighted mean for global score:
  $z_t = 0.8 \times \text{mean}\left(\left\{z_t^k\right\}\right) - 0.2 \times H(\hat{y}_i^u)$
  $z_s = 0.8 \times \text{mean}\left(\left\{z_s^k\right\}\right) - 0.2 \times H\left(\tilde{y}_i^u\right)$
  $z_b = 0.8 \times \text{mean}\left(\left\{z_b^k\right\}\right) - 0.2 \times H\left(y_i^b\right)$
  if $z_t > z_s$ and $z_t > z_b$ **then**
    Replace the y_i^b in B_U by $\hat{y}_i^u$;
  end if
end for

3.3. Multimodal Contrastive Loss

In most cases, numerous mean teacher-based approaches utilize the L1 distance as the consistency loss, as illustrated in Equation (2). The simple consistency loss may cause the student model to overfit incorrect predictions, thereby introducing confirmation bias. To alleviate this problem, we propose the integration of contrastive loss during training. Contrastive learning, a prominent approach in self-supervised learning [28], drives the model to differentiate between positive and negative sample pairs by pulling together their representations in the former case and pushing them apart in the latter. In our context, the positive samples correspond to the pseudo-labels, while the negative samples are the corresponding degraded underwater images. However, traditional contrastive learning typically performs a global comparison of features, which may be insufficient for capturing the complex characteristics of underwater images. To address this limitation, we aim to extend it into a comprehensive contrastive loss that integrates multimodal information, multi-scale feature representations, and adaptive feature selection. This enhancement is designed to significantly improve the robustness and generalization ability of the model, particularly in the generation of pseudo-labels and the effective utilization of unlabeled data.

To optimize the robustness and generalization of the student model during unsupervised learning, we introduce a multimodal contrastive loss that integrates VGG features, edge cues, color information, and local region features. This multimodal contrastive strategy enables the model to distinguish subtle variations in the image content, thereby improving pseudo-label reliability.

3.3.1. VGG Feature Contrastive Loss

Let a^i_{vgg}, p^i_{vgg}, and n^i_{vgg} represent the anchor, positive, and negative VGG features at the i-th layer, respectively. The distances are then computed as follows: anchor-positive distance: $d^i_{ap} = \| a^i_{vgg} - p^i_{vgg} \|_1$, and anchor-negative distance: $d^i_{an} = \| a^i_{vgg} - n^i_{vgg} \|_1$. The contrastive loss at layer i is:

$$L^i_{vgg} = \frac{d^i_{ap}}{d^i_{an} + \epsilon}$$

If the negative sample is harder (i.e., $d^i_{an} < d^i_{ap}$), a hard sample weight w_{hard} is applied:

$$L^i_{vgg} = L^i_{vgg} \times w_{hard}$$

To emphasize discriminative layers, we introduce static weights w_i and dynamically computed complexity-aware weights $w^i_{dynamic}$. The final VGG-based contrastive loss is as follows:

$$L_{vgg} = \sum_i w_i \times L^i_{vgg} \times w^i_{dynamic} \tag{4}$$

3.3.2. Edge Feature Contrastive Loss

Let a_{edge}, p_{edge}, and n_{edge} denote the edge features of the anchor, positive, and negative images, respectively. The loss is defined as $L_{edge} = \frac{\|a_{edge} - p_{edge}\|_1}{\|a_{edge} - n_{edge}\|_1 + \epsilon}$. If a hard negative is detected ($d^{edge}_{an} < d^{edge}_{ap}$), then apply $L_{edge} = L_{edge} \times w_{hard}$, and finally:

$$L_{edge} = L_{edge} \times w^{edge}_{dynamic} \tag{5}$$

3.3.3. Color Feature Contrastive Loss

Let a_{color}, p_{color}, and n_{color} be the color features of anchor, positive, and negative samples. We define $L_{color} = \frac{\|a_{color} - p_{color}\|_1}{\|a_{color} - n_{color}\|_1 + \epsilon}$. With hard sample adjustment ($d_{an}^{color} < d_{ap}^{color}$): $L_{color} = L_{color} \times w_{hard}$, and finally:

$$L_{color} = L_{color} \times w_{dynamic}^{color} \tag{6}$$

3.3.4. Local Region Contrastive Loss

The image is divided into four local regions. For each region j, the anchor, positive, and negative local features are a^j_{local}, p^j_{local}, and n^j_{local}. The region-wise contrastive loss is $L^j_{local} = \frac{\|a^j_{local} - p^j_{local}\|_1}{\|a^j_{local} - n^j_{local}\|_1 + \epsilon}$. If $d_{an}^{local,j} < d_{ap}^{local,j}$, apply $L^j_{local} = L^j_{local} \times w_{hard}$. The overall local contrastive loss is averaged:

$$L_{local} = \sum_{j=1}^{4} 0.25 \times L^j_{local} \tag{7}$$

The total loss combines all components:

$$L_{mcr} = L_{vgg} + L_{edge} + L_{color} + L_{local} \tag{8}$$

This comprehensive loss function enables robust and fine-grained representation learning from unlabeled underwater images, effectively suppressing confirmation bias and improving pseudo-label quality.

To enhance the adaptiveness of our contrastive loss to varying feature complexities across different layers and modalities, we introduce a dynamic weighting scheme based on feature variance. The intuition is that feature maps with higher variance often contain richer structural or semantic information, and should contribute more significantly to the contrastive learning process.

For the i-th feature layer or modality, we define the dynamic weight $w^i_{dynamic}$ as Equation (9):

$$w^i_{dynamic} = \frac{\mathrm{Var}(f^i)}{\sum_j \mathrm{Var}(f^j)}, \ \mathrm{Var}\left(f^i\right) = \frac{1}{N}\sum_{k=1}^{N}\left(f^i_k - \overline{f}^i\right)^2 \tag{9}$$

Here, f^i denotes the feature map of the i-th layer, f^i_k is the k-th pixel (or feature vector), $\overline{f}^i$ is the mean feature value, and N is the total number of pixels in the feature map. This normalization ensures that weights across all layers or modalities sum to 1, stabilizing the training. By integrating this variance-based dynamic weighting, the model can prioritize feature levels that contain more discriminative information, leading to more effective pseudo-label learning and better generalization in complex underwater environments.

Building on Equation (2), the overall training objective is composed of a supervised loss and a refined unsupervised loss, as detailed below.

To optimize the effectiveness of the supervised loss beyond the standard L1 formulation presented in Equation (2), we adopt a more comprehensive objective inspired by [18], which incorporates not only the pixel-wise L1 loss but also a perceptual loss L_{per} and a gradient penalty term L_{grad}, thereby encouraging the restoration of finer textures and more accurate structural details.

$$L'_{sup} = L_{sup} + \alpha_1 L_{per} + \alpha_2 L_{grad} \tag{10}$$

For the unsupervised component, we refine the original L_{unsup} by formulating it as a combination of the proposed reliable teacher–student consistency loss and the multimodal contrastive loss, aiming to enhance the stability of pseudo-label learning and improve the model's generalization on unlabeled data.

$$L''_{unsup} = L'_{unsup} + \beta L_{mcr} \quad (11)$$

Finally, the overall optimization objective is reformulated as follows, consistent with the structure of Equation (2):

$$L_{overall} = L'_{sup} + \rho L''_{unsup} \quad (12)$$

where L'_{sup} denotes the enhanced supervised loss incorporating perceptual and gradient components, and L''_{unsup} represents the improved unsupervised loss that combines reliable consistency and contrastive constraints.

We adopt an initial learning rate of $2 \times e^{-4}$ and train the model for 200 epochs. The learning rate is decayed by a factor of 0.1 after 100 epochs to facilitate stable convergence. During training, all images are uniformly cropped to a 256 × 256 resolution. For data augmentation, we apply standard geometric transformations (resizing, random cropping, and rotation) to the labeled data. Regarding the unlabeled data, the teacher branch receives weakly augmented inputs (resizing only), while the student branch is exposed to strong augmentations, including resizing, color jittering, Gaussian blur, and grayscale conversion, to encourage consistency under perturbations. The loss function comprises several components, whose weights are empirically set as follows: $\alpha_1 = 0.3$, $\alpha_2 = 0.1$, and $\beta = 1$. Additionally, the consistency regularization coefficient ρ is gradually increased during training using an exponential equation [37]: $\rho(t) = 0.2 \times e^{-5(1-t/200)^2}$, where t denotes the training epoch. This progressive scheduling helps stabilize training in the early stages by controlling the influence of the unsupervised loss component.

4. Experimental Results

4.1. Datasets and Settings

4.1.1. Software Configuration

The proposed approach was developed using the PyTorch (1.11.0 + cu113) framework and executed on an NVIDIA RTX 4090 D GPU. To accelerate convergence and minimize training duration, the AdamP optimizer [38] was employed, owing to its efficiency in reaching optimal solutions.

4.1.2. Introduction to Dataset

The training dataset comprises 1600 labeled image pairs and 1600 unlabeled images. The labeled pairs are randomly selected in an equal proportion from the dataset proposed in [39] and the UIEB dataset [20]. Specifically, ref. [39] provides a collection of synthetically generated underwater images captured in indoor scenes, while UIEB [20] contains 890 real-world underwater images accompanied by corresponding ground truth references. The unpaired subset of the EUVP benchmark [22] serves as the source of unlabeled images, which includes diverse underwater scenes with varying water types and illumination conditions. To evaluate performance, the test set incorporates both full-reference and no-reference benchmark datasets, comprising 89 images from the UIEB and 500 images from the LSUI [40].

4.1.3. Evaluation Metrics

To evaluate model performance, we utilize a set of commonly adopted image quality assessment metrics. In particular, peak signal-to-noise ratio (PSNR), structural similarity index (SSIM), and root mean square error (RMSE) are selected as full-reference metrics. A higher PSNR signifies superior image fidelity, whereas SSIM values approaching 1 denote stronger structural resemblance to the ground truth. Conversely, a lower RMSE implies a reduced restoration error. Additionally, two no-reference evaluation metrics—underwater image quality measure (UIQM) and underwater color image quality evaluation (UCIQE)—are applied to assess the perceptual quality of the enhanced underwater images. Elevated UIQM and UCIQE scores reflect improved visual appeal and color restoration accuracy. The calculation formulas for the evaluation metrics are presented as follows: The definitions of PSNR, SSIM, and RMSE are provided in Equations (13), (14), and (15), respectively.

$$\mathrm{PSNR} = 10log_{10}(\frac{max^2}{\mathrm{MSE}}) \tag{13}$$

$$\mathrm{SSIM} = \frac{(2\mu_x\mu_y + c_1)(2\sigma_{xy} + c_2)}{(\mu_x^2+\mu_y^2 + c_1)(\sigma_x^2+\sigma_y^2 + c_2)} \tag{14}$$

$$\mathrm{RMSE} = \sqrt{\mathrm{MSE}} = \sqrt{\frac{1}{MN}\sum_{i=1}^{M}\sum_{j=1}^{N}[y_e(i,j) - y(i,j)]^2} \tag{15}$$

UIQM is a no-reference metric designed to evaluate the perceptual quality of underwater images by integrating three aspects: colorfulness, sharpness, and contrast. The overall UIQM is calculated as a weighted combination of these components, as shown in Equation (16):

$$\mathrm{UIQM} = c_1 \cdot \mathrm{UICM} + c_2 \cdot \mathrm{UISM} + c_3 \cdot \mathrm{UIConM} \tag{16}$$

Among them, the commonly used weight coefficients are as follows: c_1 = 0.0282, c_2 = 0.2953, and c_3 = 3.5753. UCIQE is a no-reference metric used to assess the perceptual quality of underwater images. It primarily considers three visual attributes: colorfulness, contrast, and saturation. The UCIQE score is computed as a weighted linear combination of the standard deviation of chroma (ω_c), the contrast of luminance (con_l), and the mean of saturation (μ_s), as defined in Equation (17):

$$\mathrm{UCIQE} = c_1 \cdot \omega_c + c_2 \cdot con_l + c_3 \cdot \mu_s \tag{17}$$

where c_1 = 0.4680, c_2 = 0.2745, and c_3 = 0.2576 are the empirically determined weighting coefficients.

4.2. Enhanced Experiments on Public Datasets

We first conduct enhancement experiments on the UIEB test set, which contains 89 real-world underwater images, each resized to 256 × 256 pixels for evaluation. To validate the effectiveness of our approach, we compare it against several representative underwater image enhancement approaches, including both traditional and deep learning-based methods: NLD [41], CLAHE [6], DCP [42], UDCP [43], UNet [44], UWNet [45], CycleGAN [46], and FUnIE-GAN [22]. A selection of enhanced images are shown in Figure 3 for visual comparison. Specifically, (a) is the input image; (b–j) are the results of NLD, CLAHE, DCP, UDCP, UNet, UWNet, CycleGAN, FUnIE-GAN, and our MCR-UIE, respectively; and (k) is ground truth.

As shown in Figure 3, deep learning-based approaches clearly outperform traditional enhancement approaches in terms of visual quality. This superiority can be attributed to the ability of deep learning models to automatically learn complex representations from data. These models effectively capture subtle textures, edges, and color distributions in underwater scenes. In contrast, traditional approaches typically rely on hand-crafted features and heuristic adjustments, such as enhancing brightness or contrast, which are insufficient to fully recover the degraded content and intricate details present in underwater images.

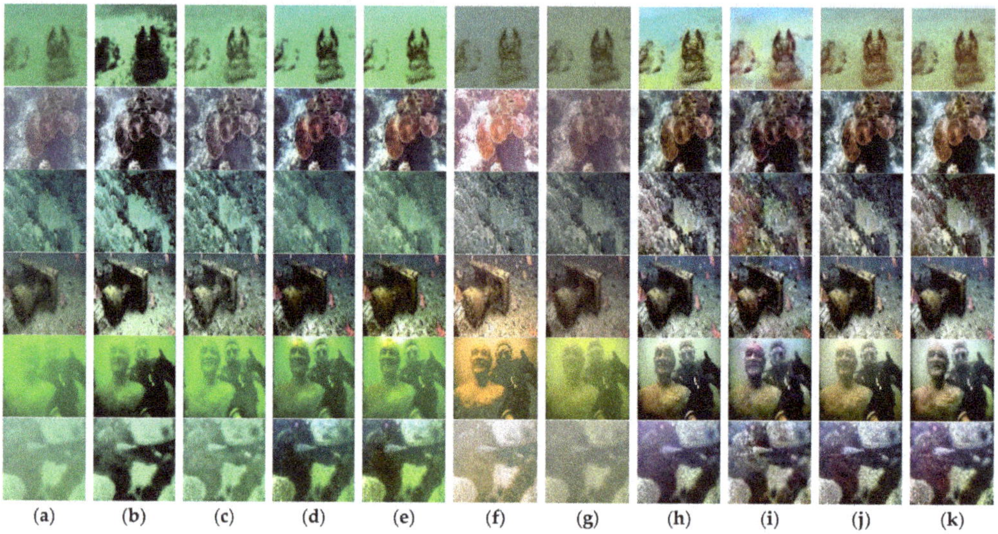

Figure 3. Visual comparison of enhancement effects on UIEB dataset. (**a**) Input; (**b**) NLD; (**c**) CLAHE; (**d**) DCP; (**e**) UDCP; (**f**) UNet; (**g**) UWNet; (**h**) CycleGAN; (**i**) FUnIE-GAN; (**j**) MCR-UIE; (**k**) ground truth.

To further demonstrate the efficacy of our proposed approach, additional enhancement experiments were performed on the LSUI dataset. The LSUI test set contains 500 underwater images, each resized to 256 × 256 pixels for evaluation. A selection of enhanced images are shown in Figure 4 for visual comparison. Specifically, (a) is the input image; and (b–j) are the outputs of NLD, CLAHE, DCP, UDCP, UNet, UWNet, CycleGAN, FUnIE-GAN, and our MCR-UIE, respectively.

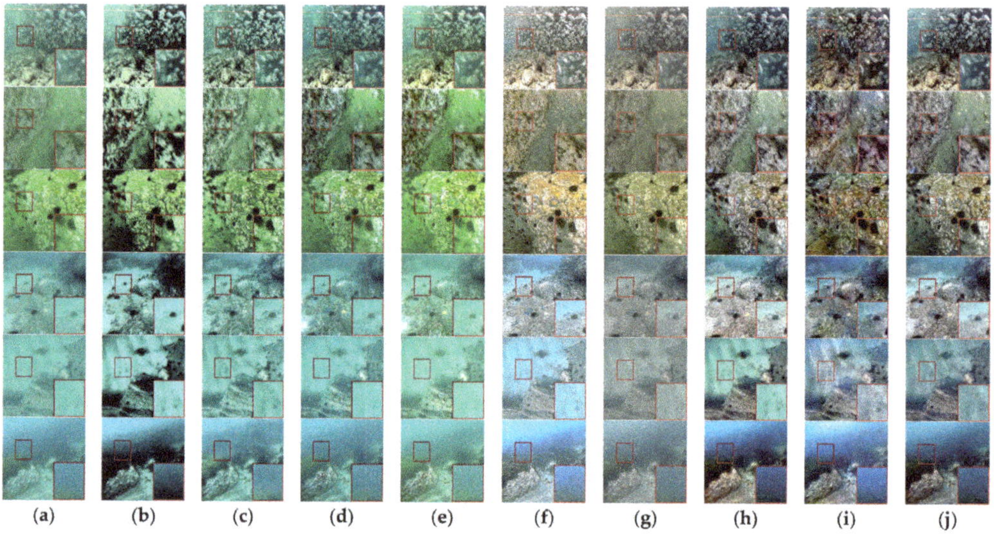

Figure 4. Visual comparison of enhancement effects on LSUI dataset. (**a**) Input; (**b**) NLD; (**c**) CLAHE; (**d**) DCP; (**e**) UDCP; (**f**) UNet; (**g**) UWNet; (**h**) CycleGAN; (**i**) FUnIE-GAN; (**j**) MCR-UIE.

To better visualize the enhancement effects, a representative region was selected and zoomed in on, presented in the lower right corner of Figure 4. Upon closer inspection, it is evident that the LSUI test set more closely resembles real underwater scenes in terms of visual characteristics. As with previous results, deep learning-based approaches continue to outperform traditional approaches. Notably, the CycleGAN method produces images with relatively high color saturation, which may lead to unrealistic results. In contrast, our proposed approach achieves superior visual performance, preserving natural color tones and fine details more effectively than the other compared methods. While subjective visual comparisons provide an intuitive understanding of enhancement quality, they are insufficient for a comprehensive evaluation. Therefore, we further assess the performance using objective evaluation metrics, which include both full-reference and no-reference metrics. The final reported numbers represent the average values across the entire test set. Table 1 presents the full-reference evaluation results on the UIEB and LSUI datasets after image enhancement, where the best and second-best results are marked in red and blue, respectively.

Table 1. Performance metrics using different enhancement approaches (PSNR, SSIM, RMSE).

Method	UIEB			LSUI		
	PSNR ↑ (dB)	SSIM ↑	RMSE ↓	PSNR ↑	SSIM ↑	RMSE ↓
NLD	16.416	0.708	41.261	14.629	0.694	49.862
CLAHE	16.812	0.751	39.182	14.713	0.744	48.154
DCP	16.526	0.713	41.558	14.025	0.694	52.976
UDCP	17.478	0.752	36.284	15.613	0.756	43.976
UNet	14.668	0.706	50.550	16.851	0.772	38.738
UWNet	17.771	0.759	36.146	18.782	0.783	31.139
CycleGAN	21.723	0.795	22.694	20.570	0.784	25.124
FUnIE-GAN	19.524	0.784	27.584	17.948	0.777	32.951
MCR-UIE	23.698	0.851	18.089	22.835	0.865	19.612

As presented in Table 1, the proposed MCR-UIE approach significantly outperforms traditional underwater image enhancement techniques, including NLD, CLAHE, DCP, and UDCP. On the UIEB dataset, MCR-UIE improves the PSNR by 44.3% and SSIM by 20.2% compared to the best-performing traditional method (UDCP). Similarly, on the LSUI dataset, MCR-UIE achieves a 46.3% increase in PSNR and a 14.4% improvement in SSIM over UDCP. Additionally, MCR-UIE reduces the RMSE by 46.0% on UIEB and 55.4% on LSUI, demonstrating its superior ability to recover image details and reduce distortion.

Compared to deep learning-based methods which encompass UNet, UWNet, CycleGAN, and FUnIE-GAN, MCR-UIE also achieves consistent performance gains. On the UIEB dataset, MCR-UIE surpasses CycleGAN in PSNR and SSIM by 9.1% and 7.0%, respectively, while reducing RMSE by 20.3%. On the LSUI dataset, MCR-UIE improves PSNR by 11.0%, SSIM by 10.3%, and lowers RMSE by 21.9% compared to CycleGAN. These results confirm that our approach not only effectively restores underwater image quality but also exhibits a strong generalization capability across different datasets.

To provide a more intuitive comparison of performance differences among the methods, Figure 5 presents the box plots of the PSNR and RMSE metrics corresponding to the results in Table 1. These plots illustrate the distribution, central tendency, and variability of each method's performance across the test set, enabling a clearer visual interpretation of their relative effectiveness.

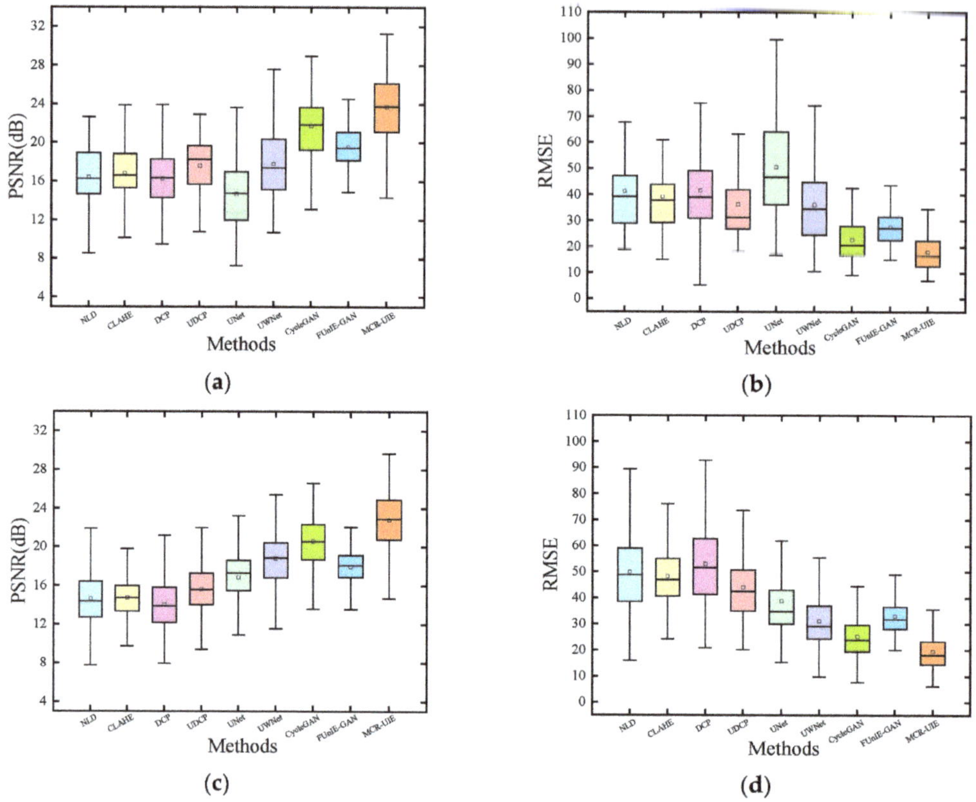

Figure 5. Box plots of performance metrics for different enhancement methods. (**a**) PSNR of UIEB, (**b**) RMSE of UIEB, (**c**) PSNR of LSUI, (**d**) RMSE of LSUI.

Table 2 reports the results of no-reference quality evaluation metrics, including UIQM and UCIQE, on the UIEB and the LSUI datasets. These metrics are designed to assess image quality in the absence of ground truth by evaluating attributes such as colorfulness, contrast, and sharpness. From the results, we observe that deep learning-based approaches generally outperform traditional enhancement approaches. Notably, UNet achieves the highest UIQM score (3.075) on the LSUI dataset, indicating its strong ability to enhance image contrast and sharpness in certain scenes. However, its UCIQE performance remains relatively modest (0.532), suggesting potential issues in maintaining consistent color balance.

Our proposed MCR-UIE method obtains a UIQM of 2.881 on UIEB and 3.000 on LSUI, ranking among the top-performing methods across both datasets. Although its UIQM is slightly lower than that of UNet and FUnIE-GAN, MCR-UIE shows a more stable and balanced performance, with UCIQE scores of 0.606 (UIEB) and 0.572 (LSUI) that are consistently high across datasets. In contrast, some methods (e.g., CycleGAN and UWNet) exhibit strong UIQM but relatively poor UCIQE, indicating possible color over-enhancement or inconsistency.

In summary, MCR-UIE achieves a strong trade-off between sharpness, color fidelity, and contrast, producing visually pleasing results while maintaining generalization ability across different underwater environments. This is also consistent with the qualitative results shown in Figure 4. To provide a more intuitive comparison of the results presented in

Table 2, Figure 6 shows a bar chart to highlight the differences in image quality enhancement across different metrics.

Table 2. Performance metrics using different enhancement methods (UIQM, UCIQE). The best and second-best results are marked in red and blue, respectively.

Method	UIQM ↑		UCIQE ↑	
	UIEB	LSUI	UIEB	LSUI
NLD	2.518	2.540	0.600	0.571
CLAHE	2.665	2.515	0.562	0.523
DCP	2.386	2.410	0.602	0.558
UDCP	2.829	2.821	0.601	0.559
UNet	2.810	3.075	0.573	0.532
UWNet	2.849	2.905	0.531	0.498
CycleGAN	2.850	2.997	0.604	0.508
FUnIE-GAN	3.033	3.069	0.614	0.586
MCR-UIE	2.881	3.000	0.606	0.572

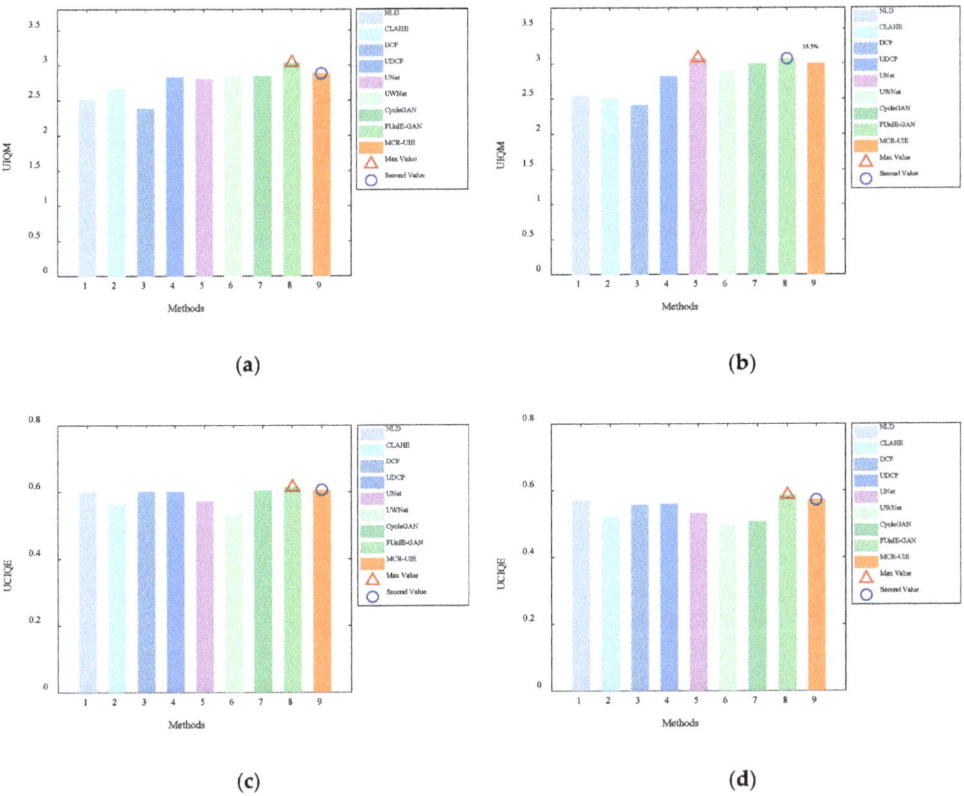

Figure 6. Bar charts of performance metrics for different enhancement methods. (**a**) UIQM of UIEB, (**b**) UIQM of LSUI, (**c**) UCIQE of UIEB, (**d**) UCIQE of LSUI.

To comprehensively evaluate the performance of different underwater image enhancement methods, we analyze both full-reference metrics (Table 1) and no-reference metrics (Table 2) across the UIEB and LSUI datasets. From the perspective of full-reference metrics including PSNR, SSIM, and RMSE, our proposed MCR-UIE achieves the best overall performance.

In terms of no-reference evaluation metrics, MCR-UIE also performs competitively. However, it is worth noting that although FUnIE-GAN performs well on no-reference metrics, its performance on full-reference metrics (e.g., PSNR = 19.524 dB on UIEB) is relatively low, suggesting that its visual enhancement may not be structurally accurate. In contrast, MCR-UIE achieves a strong balance between full-reference and no-reference metrics, with consistently high UIQM (2.881/3.000) and UCIQE (0.606/0.572) values across both datasets. This indicates that MCR-UIE not only preserves structural fidelity but also enhances visual perception quality effectively.

Overall, the proposed MCR-UIE framework exhibits excellent generalization, stable enhancement quality, and balanced performance from both subjective and objective evaluation perspectives, surpassing traditional and recent deep learning-based underwater image enhancement methods. On widely recognized public benchmarks such as UIEB and LSUI, our method achieves notable improvements in visual clarity, color fidelity, and contrast restoration, validating its effectiveness under diverse underwater conditions and degradation types. These results confirm that the proposed multimodal contrastive regularization and dynamic quality reliability strategy are highly effective in guiding the network toward producing perceptually pleasing and semantically faithful outputs.

Given its strong performance on benchmark datasets and its robustness in real-world scenarios, MCR-UIE holds great promise for practical deployment in underwater visual applications, especially in aquaculture environments. Specifically, as detailed in Section 4.3, we conduct underwater image enhancement experiments using images collected from deep-sea aquaculture cages. The model demonstrates a strong adaptability and enhancement capability in complex oceanic environments, characterized by low visibility, high turbidity, and dynamic illumination. These experiments not only confirm the real-world utility of MCR-UIE but also pave the way for its integration into intelligent monitoring systems in aquaculture, such as fish detection, behavior analysis, and health condition assessment.

4.3. Enhanced Experiments on Deep-Sea Cage Dataset

To further validate the effectiveness and generalization ability of the proposed MCR-UIE method in real-world scenarios, we deployed our model in the context of the deep-sea cage aquaculture, where underwater images are typically degraded due to the low illumination, high turbidity, and non-uniform color distortion caused by complex and dynamic oceanic environments.

We conducted two sets of underwater image enhancement experiments using images captured from the same large-scale deep-sea aquaculture cage on 16 April and 17 April 2025, respectively. These images were taken during the routine monitoring of Trachinotus ovatus in deep-sea cages and present significant visual challenges that conventional enhancement methods struggle to address.

The first test set, collected on 16 April, consisted of 350 underwater images mainly characterized by a bluish-green color cast. Four representative images from this test set were selected to visually compare the enhancement results before and after applying MCR-UIE, as illustrated in Figure 7. The second test set, collected on 17 April, included 380 images with a yellowish-green background. Similarly, four enhanced image samples were chosen from this set to show the improvements achieved by our model, as demonstrated in Figure 8. These color differences reflect environmental variations such as water turbidity, light penetration, and biological activity.

We employed two widely used no-reference image quality metrics, UIQM and UCIQE, to assess the visual quality of the degraded input images. The first test set yielded an average UIQM score of 1.810 and a UCIQE score of 0.456. The second test set achieved slightly higher values, with an average UIQM of 1.905 and UCIQE of 0.485.

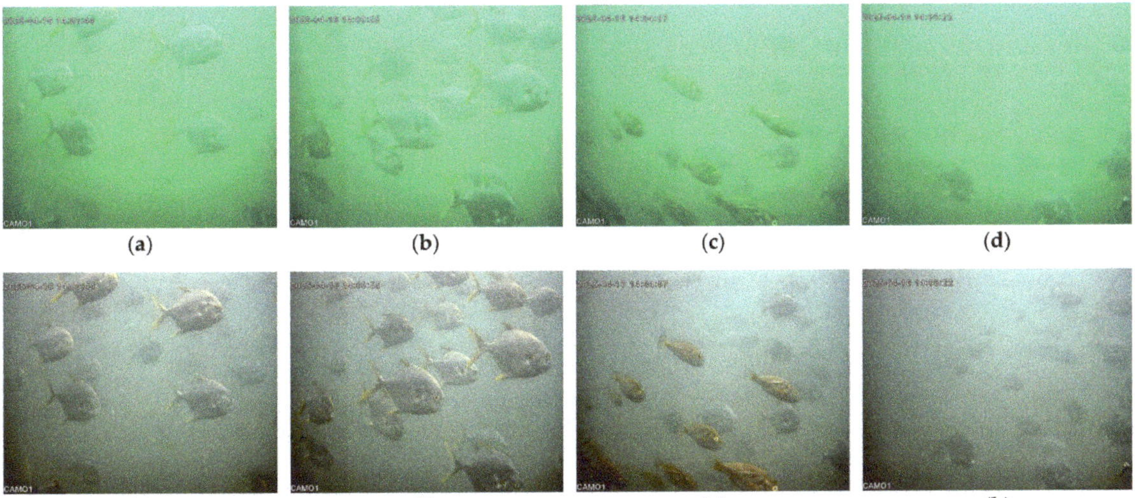

(a) (b) (c) (d)

(e) (f) (g) (h)

Figure 7. Example 1 of using MCR-UIE to enhance underwater images of deep-sea cage aquaculture. (**a**–**d**) Raw image captured in aquaculture environment; (**e**–**h**) enhanced output by our method.

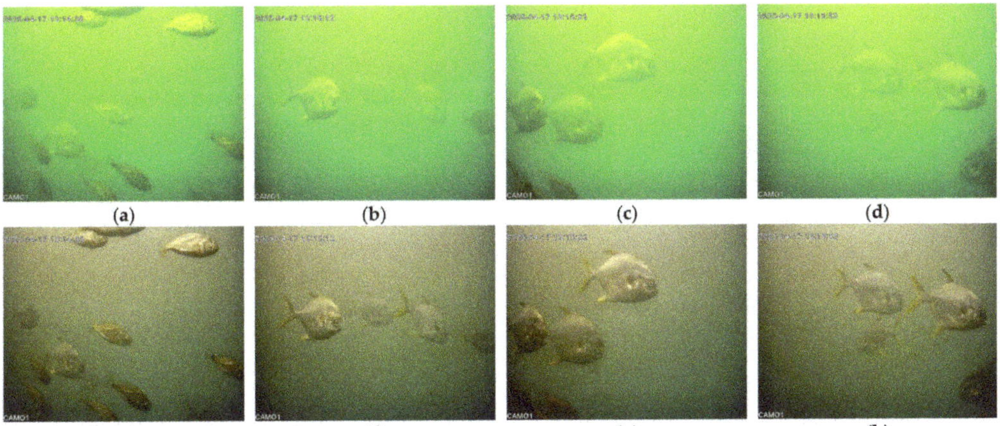

(a) (b) (c) (d)

(e) (f) (g) (h)

Figure 8. Example 2 of using MCR-UIE to enhance underwater images of deep-sea cage aquaculture. (**a**–**d**) Raw image captured in aquaculture environment; (**e**–**h**) enhanced output by our method.

Interestingly, while the second test set yielded higher objective scores, subjective evaluation suggested that the first test set produced more visually pleasing results, with clearer textures, more natural color reproduction, and better overall visual appeal. This discrepancy suggests that current NR-IQA metrics may not fully capture perceptual quality in complex underwater scenes. For instance, the yellowish-green cast in the second set may have led to artificially higher metric scores due to increased global contrast or saturation, even though human observers preferred the more naturally enhanced results from the first set.

These findings highlight the limitations of relying solely on objective metrics for underwater image evaluation. They also emphasize the need to integrate both quantitative and qualitative assessments when validating enhancement models in practical deployments.

In summary, the results from these two real-world test sets demonstrate that MCR-UIE can significantly improve the visual quality of underwater images captured under varying environmental conditions in deep-sea aquaculture. The model effectively suppresses color casts, enhances texture details, and boosts the overall image contrast. This confirms its practical potential for deployment in intelligent aquaculture monitoring systems, supporting downstream tasks such as fish detection, length measurement, and behavioral analysis.

However, we acknowledge that the current evaluation in practical deep-sea aquaculture scenarios primarily relies on qualitative comparisons and NR-IQA metrics, due to the absence of paired ground-truth images. Acquiring such reference data in real-world underwater environments is inherently challenging because of dynamic lighting conditions, uncontrollable turbidity, and the non-rigid nature of underwater scenes.

Moving forward, potential strategies may include the use of synthetic datasets that simulate the degradation characteristics of deep-sea cages, or indirect validation through downstream tasks such as fish detection, tracking, and body length estimation. Additionally, expert visual assessments can provide a complementary perspective when reference data are unavailable. These approaches could help mitigate the limitations of no-reference evaluation and enhance the robustness of model validation in real-world aquaculture applications.

4.4. Ablation Experiments

To evaluate the effectiveness of MCR-UIR, we perform a series of ablation experiments to investigate the contributions of its key components. The following variants are examined: (a) Semi-base: A baseline semi-supervised framework employing the consistency loss. (b) Semi-base + DQR: Extends the Semi-base model by incorporating the dynamic quality reliability repository, while excluding multimodal contrastive loss. (c) Semi-base + MCL_1: Adds the VGG feature contrastive loss to the Semi-base model, without utilizing the DQR repository. (d) Semi-base + MCL_2: Adds the edge feature contrastive loss to the Semi-base model, without utilizing the DQR repository. (e) Semi-base + MCL_3: Adds the color feature contrastive loss to the Semi-base model, without utilizing the DQR repository. (f) Semi-base + MCL_4: Adds the local region contrastive loss to the Semi-base model, without utilizing the DQR repository. (g) Semi-base + MCL: Adds the multimodal contrastive loss to the Semi-base model, without utilizing the DQR repository. (h) MCR-UIR: The complete proposed method, integrating both the DQR repository and the multimodal contrastive loss.

The qualitative comparisons are illustrated in Figure 9, with particular emphasis on the results of Semi-base + MCL and Semi-base + DQR. In addition, the quantitative results are given in Table 3.

(1) For Semi-base + MCL, due to the absence of reliable positive samples, the contrastive loss compels the network to differentiate excessively from the negative samples (i.e., the input images), which unfortunately leads to over-enhancement artifacts.

(2) To further dissect the impact of each modality within the multimodal contrastive loss, we additionally conduct experiments using individual contrastive branches, including VGG feature contrastive loss (MCL_1), edge feature contrastive loss (MCL_2), color feature contrastive loss (MCL_3), and local region contrastive loss (MCL_4). The results show that among the single-modality variants, the MCL_3 achieves the most competitive performance, particularly on the LSUI dataset, suggesting its strong contribution to real-world color restoration. Conversely, the MCL_4 leads to relatively poor performance, possibly due to its sensitivity to local distortions and instability in degraded underwater scenes. When combining all four modalities into a unified MCL framework, we observe consistent improvements across both datasets. The full MCL configuration, which integrates all modalities, achieves better results than any single component, confirming the complementary benefits of multimodal representations.

(3) In contrast, Semi-base + DQR lacks the contrastive regularization mechanism, and although it benefits from the dynamic quality reliability repository, the restored images still exhibit noticeable color distortions and remain visually similar to the degraded inputs.

These observations validate the complementary effectiveness of both the dynamic quality reliability repository and multimodal contrastive regularization in improving restoration quality.

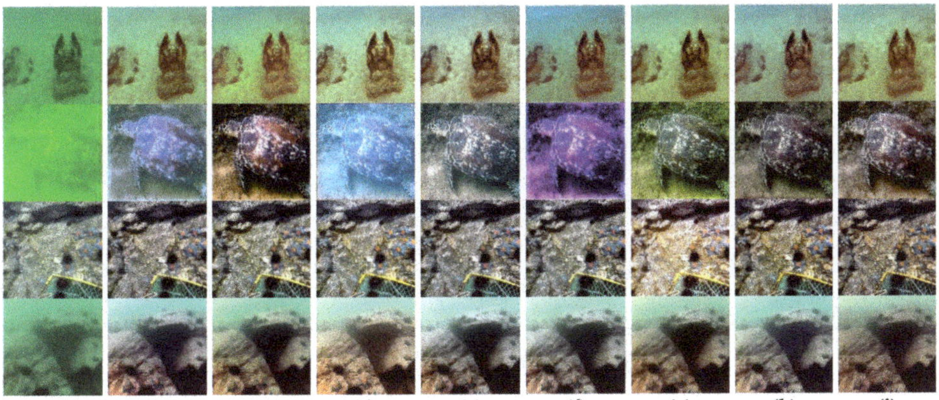

(a) (b) (c) (d) (e) (f) (g) (h) (i)

Figure 9. Visual comparison of ablation results using representative images from the UIEB and LSUI datasets. (**a**) Input; (**b**) Semi-base; (**c**) Semi-base + DQR; (**d**) Semi-base + MCL_1; (**e**) Semi-base + MCL_2; (**f**) Semi-base + MCL_3; (**g**) Semi-base + MCL_4; (**h**) Semi-base + MCL; (**i**) MCR-UIE.

Table 3. Ablation studies on UIEB and LSUI datasets with PSNR, SSIM, and RMSE. Bold metrics represent the best results.

Method	UIEB			LSUI		
	PSNR ↑	SSIM ↑	RMSE ↓	PSNR ↑	SSIM ↑	RMSE ↓
Semi-base	22.985	0.847	19.503	21.982	0.850	22.003
Semi-base + DQR	23.201	0.848	19.013	22.285	0.861	20.892
Semi-base + MCL_1	21.902	0.837	22.205	21.165	0.785	26.121
Semi-base + MCL_2	22.282	0.840	21.145	21.784	0.845	22.309
Semi-base + MCL_3	22.562	0.844	20.355	22.030	0.853	21.817
Semi-base + MCL_4	21.898	0.836	22.670	19.826	0.835	30.048
Semi-base + MCL	22.759	0.838	20.977	22.356	0.849	23.151
MCR-UIE	**23.698**	**0.851**	**18.089**	**22.835**	**0.865**	**19.612**

4.5. Deployment Feasibility

To assess the practical deployment potential of the proposed MCR-UIE model, we evaluate its inference speed on two underwater image enhancement benchmarks used in this experiment. The model contains approximately 1.68 million parameters (1,675,281) and is designed to be lightweight and computationally efficient.

We first conduct inference tests on the UIEB dataset, which comprises 89 underwater images with a resolution of 256 × 256 pixels. The total inference time on an NVIDIA RTX 4090D GPU is 2.827 s, yielding an average inference time of 31.76 ms per image, corresponding to approximately 31.48 frames per second (FPS). Additionally, on the LSUI test set, which includes 500 images of the same resolution, the total inference time is 13.106 s, resulting in an average of 26.20 ms per image, or about 38.15 FPS.

These results demonstrate that the proposed method achieves near real-time performance, indicating its strong potential for real-world applications that require timely underwater image enhancement. In particular, such efficiency suggests its practical applicability in aquaculture monitoring systems, where high-throughput and low-latency visual processing are essential.

4.6. Analysis of Limitations

To provide a more comprehensive understanding of the limitations of the proposed MCR-UIE method, we present three representative cases, as shown in Figure 10. These examples illustrate how the model may underperform or introduce perceptual distortions under specific conditions.

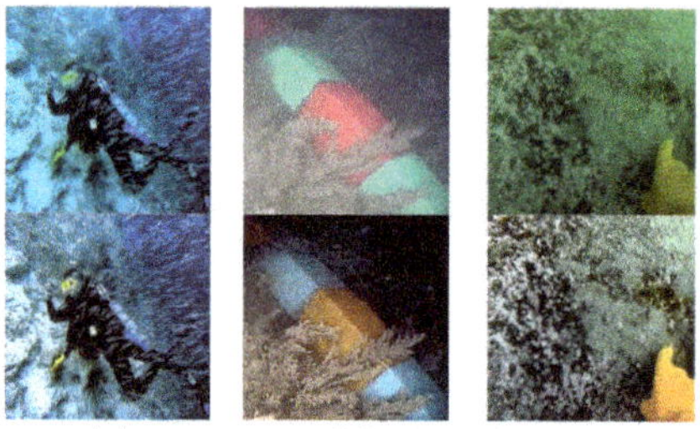

(a) (b) (c)

Figure 10. Challenging cases that reflect the limitations of MCR-UIE. (**a**) Example 1; (**b**) example 2; (**c**) example 3.

Figure 10a This image features a high-resolution underwater scene with a clear blue background and a large number of fish. After enhancement, the output shows negligible changes. This suggests that the model is overly conservative in scenes that deviate from the typical degradation patterns present in the training data. The low degree of visible enhancement may result from the pseudo-label filtering mechanism and the consistency constraint, which prevent aggressive adjustments when degradation is not apparent. However, in practical applications, such scenes may still benefit from subtle contrast or clarity improvements, indicating a gap between perceptual enhancement and model behavior.

Figure 10b The input contains an object with alternating green and red bands—an uncommon color pattern in typical underwater datasets. After enhancement, these colors shift to blue and orange. This semantic color distortion reflects the model's difficulty in preserving object colors that fall outside its learned color priors. Since the model is trained primarily on degraded natural underwater scenes without explicit object or semantic supervision, it may misinterpret such color combinations as artifacts and apply misguided corrections, leading to visually unrealistic results.

Figure 10c In this case, the input image includes a bright yellow rubber glove, which is transformed to have an orange hue after enhancement. The color shift suggests that the model mistakenly treats vivid artificial objects as being affected by underwater color distortion. Without an understanding of object semantics or color constancy, the enhancement process adjusts the hue toward what it believes is a more "natural" underwater tone, resulting in a loss of color fidelity for human-made objects.

These cases highlight two major limitations: a lack of semantic understanding, which causes the model to misinterpret unusual colors as degradations and incorrectly adjust them, and limited adaptability to high-quality or complex scenes, where the model's enhancement behavior becomes conservative or misaligned with perceptual needs.

To address these issues, future work could consider integrating semantic segmentation priors, human-object-aware loss functions, or expanding the training dataset to include a greater diversity of object colors and scene types. In particular, increasing the diversity of the unlabeled dataset, which currently consists solely of EUVP samples, would mitigate potential domain shift effects and allow the model to better adapt to underwater environments not represented in the original benchmark. This expanded dataset could incorporate varied water types, lighting conditions, and object appearances, thereby helping the model distinguish between actual degradation and semantically meaningful color variations. Together, these enhancements would improve the model's robustness and generalization for real-world deployment scenarios such as deep-sea aquaculture monitoring.

5. Conclusions

In this study, we propose a novel semi-supervised underwater image enhancement approach, termed MCR-UIE, which integrates a multimodal contrastive learning (MCL) and a dynamic quality reliability repository (DQR) to fully exploit both labeled and unlabeled data. The MCL introduces contrastive constraints across multiple modalities, such as perceptual, edge, color, and local region features, to enhance feature discrimination and robustness. Meanwhile, the DQR maintains high-quality pseudo-labels by dynamically selecting the best teacher outputs based on a learned reliability criterion, thus mitigating the risk of confirmation bias during training. A wide range of experiments conducted on multiple underwater image enhancement benchmarks demonstrate that MCR-UIE consistently outperforms existing advanced methods, achieving notable improvements across both full-reference and no-reference evaluation metrics. In subsequent research, we plan to extend this semi-supervised framework to other low-level vision tasks and explore more efficient memory-aware training strategies to further improve scalability and performance.

Author Contributions: Conceptualization, M.D. and G.L.; methodology, M.D.; software, M.D.; validation, G.L. and H.L.; formal analysis, Q.H.; writing—original draft preparation, M.D.; writing—review and editing, M.D. and X.H.; visualization, Y.H.; supervision, Q.H.; project administration, Y.H. and X.H.; funding acquisition, M.D., G.L., H.L. and X.H. All authors have read and agreed to the published version of the manuscript.

Funding: This research was funded by the earmarked fund for CARS-47, the Central Public-interest Scientific Institution Basal Research Fund, CAFS (No. 2023TD97), and the Central Public-interest Scientific Institution Basal Research Fund, South China Sea Fisheries Research Institute, CAFS (No. 2023RC01, No. 2022TS06, No. 2024TS07 and No. 2024TS08).

Data Availability Statement: The datasets used in this study are available from the corresponding author upon reasonable request. They are not publicly released in order to prevent potential misuse.

Conflicts of Interest: The authors declare no conflicts of interest.

References

1. Shuang, X.; Zhang, J.; Tian, Y. Algorithms for improving the quality of underwater optical images: A comprehensive review. *Signal Process.* **2024**, *219*, 109408. [CrossRef]
2. Rout, D.K.; Kapoor, M.; Subudhi, B.N.; Thangaraj, V.; Jakhetiya, V.; Bansal, A. Underwater visual surveillance: A comprehensive survey. *Ocean Eng.* **2024**, *309*, 118367. [CrossRef]
3. Berman, D.; Levy, D.; Avidan, S.; Treibitz, T. Underwater single image color restoration using haze-lines and a new quantitative dataset. *IEEE Trans. Pattern Anal. Mach. Intell.* **2020**, *43*, 2822–2837. [CrossRef]
4. Galdran, A.; Pardo, D.; Picón, A.; Alvarez-Gila, A. Automatic red-channel underwater image restoration. *J. Vis. Commun. Image Represent.* **2015**, *26*, 132–145. [CrossRef]
5. Zhang, S.; Wang, T.; Dong, J.; Yu, H. Underwater image enhancement via extended multi-scale Retinex. *Neurocomputing* **2017**, *245*, 1–9. [CrossRef]
6. Zuiderveld, K.J. Contrast limited adaptive histogram equalization. *Graph. Gems* **1994**, *4*, 474–485.
7. Wen, Z.; Zhao, Y.; Gao, F.; Su, H.; Rao, Y.; Dong, J. NUAM-Net: A Novel Underwater Image Enhancement Attention Mechanism Network. *J. Mar. Sci. Eng.* **2024**, *12*, 1216. [CrossRef]

8. Zhang, B.; Fang, J.; Li, Y.; Wang, Y.; Zhou, Q.; Wang, X. GFRENet: An Efficient Network for Underwater Image Enhancement with Gated Linear Units and Fast Fourier Convolution. *J. Mar. Sci. Eng.* **2024**, *12*, 1175. [CrossRef]
9. Zhao, S.; Mei, X.; Ye, X.; Guo, S. MSFE-UIENet: A Multi-Scale Feature Extraction Network for Marine Underwater Image Enhancement. *J. Mar. Sci. Eng.* **2024**, *12*, 1472. [CrossRef]
10. Yang, J.; Zhu, S.; Liang, H.; Bai, S.; Jiang, F.; Hussain, A. PAFPT: Progressive aggregator with feature prompted transformer for underwater image enhancement. *Expert Syst. Appl.* **2025**, *262*, 125539. [CrossRef]
11. Xiang, D.; He, D.; Sun, H.; Gao, P.; Zhang, J.; Ling, J. HCMPE-Net: An unsupervised network for underwater image restoration with multi-parameter estimation based on homology constraint. *Opt. Laser Technol.* **2025**, *186*, 112616. [CrossRef]
12. Fu, F.; Liu, P.; Shao, Z.; Xu, J.; Fang, M.M.-G. A multi-scale evolutionary generative adversarial network for underwater image enhancement. *J. Mar. Sci. Eng.* **2024**, *12*, 1210. [CrossRef]
13. Wang, M.; Zhang, K.; Wei, H.; Chen, W.; Zhao, T. Underwater image quality optimization: Researches, challenges, and future trends. *Image Vis. Comput.* **2024**, *146*, 104995. [CrossRef]
14. Tarvainen, A.; Valpola, H. Mean teachers are better role models: Weight-averaged consistency targets improve semi-supervised deep learning results. In Proceedings of the 31st International Conference on Neural Information Processing Systems, Long Beach, CA, USA, 4–9 December 2017.
15. Huang, S.; Wang, K.; Liu, H.; Chen, J.; Li, Y. Contrastive semi-supervised learning for underwater image restoration via reliable bank. In Proceedings of the IEEE/CVF Conference on Computer Vision and Pattern Recognition, Vancouver, BC, Canada, 17–24 June 2023; pp. 18145–18155.
16. Ancuti, C.O.; Ancuti, C.; De Vleeschouwer, C.; Bekaert, P. Color balance and fusion for underwater image enhancement. *IEEE Trans. Image Process.* **2017**, *27*, 379–393. [CrossRef]
17. Zhang, W.; Zhuang, P.; Sun, H.-H.; Li, G.; Kwong, S.; Li, C. Underwater image enhancement via minimal color loss and locally adaptive contrast enhancement. *IEEE Trans. Image Process.* **2022**, *31*, 3997–4010. [CrossRef]
18. Kar, A.; Dhara, S.K.; Sen, D.; Biswas, P.K. Zero-shot single image restoration through controlled perturbation of koschmieder's model. In Proceedings of the IEEE/CVF Conference on Computer Vision and Pattern Recognition, Nashville, TN, USA, 20–25 June 2021; pp. 16205–16215.
19. Wang, K.; Hu, Y.; Chen, J.; Wu, X.; Zhao, X.; Li, Y. Underwater image restoration based on a parallel convolutional neural network. *Remote Sens.* **2019**, *11*, 1591. [CrossRef]
20. Li, C.; Guo, C.; Ren, W.; Cong, R.; Hou, J.; Kwong, S.; Tao, D. An underwater image enhancement benchmark dataset and beyond. *IEEE Trans. Image Process.* **2019**, *29*, 4376–4389. [CrossRef]
21. Li, C.; Anwar, S.; Hou, J.; Cong, R.; Guo, C.; Ren, W. Underwater image enhancement via medium transmission-guided multi-color space embedding. *IEEE Trans. Image Process.* **2021**, *30*, 4985–5000. [CrossRef]
22. Islam, M.J.; Xia, Y.; Sattar, J. Fast underwater image enhancement for improved visual perception. *IEEE Robot. Autom. Lett.* **2020**, *5*, 3227–3234. [CrossRef]
23. Miyato, T.; Maeda, S.-i.; Koyama, M.; Ishii, S. Virtual adversarial training: A regularization method for supervised and semi-supervised learning. *IEEE Trans. Pattern Anal. Mach. Intell.* **2018**, *41*, 1979–1993. [CrossRef]
24. Sohn, K.; Berthelot, D.; Carlini, N.; Zhang, Z.; Zhang, H.; Raffel, C.A.; Cubuk, E.D.; Kurakin, A.; Li, C.-L. Fixmatch: Simplifying semi-supervised learning with consistency and confidence. *Adv. Neural Inf. Process. Syst.* **2020**, *33*, 596–608.
25. Wang, Y.; Wang, H.; Shen, Y.; Fei, J.; Li, W.; Jin, G.; Wu, L.; Zhao, R.; Le, X. Semi-supervised semantic segmentation using unreliable pseudo-labels. In Proceedings of the IEEE/CVF Conference on Computer Vision and Pattern Recognition, New Orleans, LA, USA, 18–24 June 2022; pp. 4248–4257.
26. Wang, L.; Yoon, K.-J. Semi-supervised student-teacher learning for single image super-resolution. *Pattern Recognit.* **2022**, *121*, 108206. [CrossRef]
27. Zhu, H.; Han, X.; Tao, Y. Semi-supervised advancement of underwater visual quality. *Meas. Sci. Technol.* **2020**, *32*, 015404. [CrossRef]
28. He, K.; Fan, H.; Wu, Y.; Xie, S.; Girshick, R. Momentum contrast for unsupervised visual representation learning. In Proceedings of the IEEE/CVF Conference on Computer Vision and Pattern Recognition, Seattle, WA, USA, 13–19 June 2020; pp. 9729–9738.
29. Liang, D.; Li, L.; Wei, M.; Yang, S.; Zhang, L.; Yang, W.; Du, Y.; Zhou, H. Semantically contrastive learning for low-light image enhancement. In Proceedings of the AAAI Conference on Artificial Intelligence, Online, 22 February–1 March 2022; pp. 1555–1563.
30. Wu, H.; Qu, Y.; Lin, S.; Zhou, J.; Qiao, R.; Zhang, Z.; Xie, Y.; Ma, L. Contrastive learning for compact single image dehazing. In Proceedings of the IEEE/CVF Conference on Computer Vision and Pattern Recognition, Nashville, TN, USA, 20–25 June 2021; pp. 10551–10560.
31. Simonyan, K.; Zisserman, A. Very deep convolutional networks for large-scale image recognition. *arXiv* **2014**, arXiv:1409.1556.
32. Han, J.; Shoeiby, M.; Malthus, T.; Botha, E.; Anstee, J.; Anwar, S.; Wei, R.; Armin, M.A.; Li, H.; Petersson, L. Underwater image restoration via contrastive learning and a real-world dataset. *Remote Sens.* **2022**, *14*, 4297. [CrossRef]

33. Polyak, B.T.; Juditsky, A.B. Acceleration of stochastic approximation by averaging. *SIAM J. Control Optim.* **1992**, *30*, 838–855. [CrossRef]
34. Yang, M.; Sowmya, A. An underwater color image quality evaluation metric. *IEEE Trans. Image Process.* **2015**, *24*, 6062–6071. [CrossRef]
35. Panetta, K.; Gao, C.; Agaian, S. Human-visual-system-inspired underwater image quality measures. *IEEE J. Ocean. Eng.* **2015**, *41*, 541–551. [CrossRef]
36. Ke, J.; Wang, Q.; Wang, Y.; Milanfar, P.; Yang, F. Musiq: Multi-scale image quality transformer. In Proceedings of the IEEE/CVF International Conference on Computer Vision, Montreal, QC, Canada, 10–17 October 2021; pp. 5148–5157.
37. Liu, Y.; Zhu, L.; Pei, S.; Fu, H.; Qin, J.; Zhang, Q.; Wan, L.; Feng, W. From synthetic to real: Image dehazing collaborating with unlabeled real data. In Proceedings of the 29th ACM International Conference on Multimedia, Chengdu, China, 20–24 October 2021; pp. 50–58.
38. Heo, B.; Chun, S.; Oh, S.J.; Han, D.; Yun, S.; Kim, G.; Uh, Y.; Ha, J.-W. Adamp: Slowing down the slowdown for momentum optimizers on scale-invariant weights. *arXiv* **2020**, arXiv:2006.08217.
39. Li, C.; Anwar, S.; Porikli, F. Underwater scene prior inspired deep underwater image and video enhancement. *Pattern Recognit.* **2020**, *98*, 107038. [CrossRef]
40. Peng, L.; Zhu, C.; Bian, L. U-shape transformer for underwater image enhancement. *IEEE Trans. Image Process.* **2023**, *32*, 3066–3079. [CrossRef] [PubMed]
41. Berman, D.; Avidan, S. Non-local image dehazing. In Proceedings of the IEEE Conference on Computer Vision and Pattern Recognition, Las Vegas, NV, USA, 27–30 June 2016; pp. 1674–1682.
42. He, K.; Sun, J.; Tang, X. Single image haze removal using dark channel prior. *IEEE Trans. Pattern Anal. Mach. Intell.* **2010**, *33*, 2341–2353. [PubMed]
43. Drews, P.; Nascimento, E.; Moraes, F.; Botelho, S.; Campos, M. Transmission estimation in underwater single images. In Proceedings of the IEEE International Conference on Computer Vision Workshops, Sydney, NSW, Australia, 2–8 December 2013; pp. 825–830.
44. Ronneberger, O.; Fischer, P.; Brox, T. U-net: Convolutional networks for biomedical image segmentation. In Proceedings of the Medical Image Computing and Computer Assisted Intervention, Munich, Germany, 5–9 October 2015; pp. 234–241.
45. Naik, A.; Swarnakar, A.; Mittal, K. Shallow-uwnet: Compressed model for underwater image enhancement (student abstract). In Proceedings of the AAAI Conference on Artificial Intelligence, Online, 2–9 February 2021; pp. 15853–15854.
46. Zhu, J.-Y.; Park, T.; Isola, P.; Efros, A.A. Unpaired image-to-image translation using cycle-consistent adversarial networks. In Proceedings of the IEEE International Conference on Computer Vision, Venice, Italy, 22–29 October 2017; pp. 2223–2232.

Journal of
*Marine Science
and Engineering*

MDPI

Article

An Improved DeepSORT-Based Model for Multi-Target Tracking of Underwater Fish

Shengnan Liu [1,2], Jiapeng Zhang [1], Haojun Zheng [1,2], Cheng Qian [1] and Shijing Liu [1,3,4,*]

1 Fishery Machinery and Instrument Research Institute, Chinese Academy of Fishery Sciences, Shanghai 200092, China; liushengnan1124@163.com (S.L.)
2 School of Navigation and Naval Architecture, Dalian Ocean University, Dalian 116023, China
3 Sanya Oceanographic Institution, Ocean University of China, Sanya 572011, China
4 East China Sea Fishery Research Institute, Chinese Academy of Fishery Sciences, Shanghai 200090, China
* Correspondence: liushijing@fmiri.ac.cn

Abstract

Precise identification and quantification of fish movement states are of significant importance for conducting fish behavior research and guiding aquaculture production, with object tracking serving as a key technical approach for achieving behavioral quantification. The traditional DeepSORT algorithm has been widely applied to object tracking tasks; however, in practical aquaculture environments, high-density cultured fish exhibit visual characteristics such as similar textural features and frequent occlusions, leading to high misidentification rates and frequent ID switching during the tracking process. This study proposes an underwater fish object tracking method based on the improved DeepSORT algorithm, utilizing ResNet as the backbone network, embedding Deformable Convolutional Networks v2 to enhance adaptive receptive field capabilities, introducing Triplet Loss function to improve discrimination ability among similar fish, and integrating Convolutional Block Attention Module to enhance key feature learning. Finally, by combining the aforementioned improvement modules, the ReID feature extraction network was redesigned and optimized. Experimental results demonstrate that the improved algorithm significantly enhances tracking performance under frequent occlusion conditions, with the MOTA metric improving from 64.26% to 66.93% and the IDF1 metric improving from 53.73% to 63.70% compared to the baseline algorithm, providing more reliable technical support for underwater fish behavior analysis.

Keywords: fish behavior; multi-object tracking; DeepSORT; low temperature stress

1. Introduction

As global marine fishery resources become increasingly strained, aquaculture has emerged as a crucial pathway for ensuring global food security. China stands out as the only major fishing nation where aquaculture production surpasses that of capture fisheries, playing a key role in alleviating pressure on global fishery resources [1]. Aquaculture products play a prominent role in improving residents' dietary structure and providing high-quality protein [2], making significant contributions to ensuring national food security. With the expansion of aquaculture scale and continuous technological advancement, aquaculture vessels have emerged as an important cultivation mode for deep-sea and offshore aquaculture operations. When aquaculture vessels navigate between different sea areas or water layers, they encounter variations in seawater temperature, including temperature differences between water layers and fluctuations in ocean currents, which result in excessively high or low water temperatures in certain regions [3,4]. The leopard coral grouper (*Plectropomus leopardus*) is a high-value tropical marine fish species that is sensitive to water temperature and environmental changes, making it susceptible to low temperature stress and resulting in abnormal behaviors. These abnormal behaviors initially manifest as surface breathing and swimming imbalance. If these precursory behaviors are not detected promptly, they tend to trigger mass mortality of fish, thereby severely affecting aquaculture efficiency and production stability [5,6].

Therefore, accurate identification of abnormal fish behaviors is of significant importance. Current primary technological approaches for detecting fish behavior include optical [7], acoustic [8], and sensor technologies [9]. Acoustic sensors are extensively uti-lized in large-scale offshore aquaculture scenarios; however, due to their acoustic propa-gation characteristics, they cannot meet the requirements for precise observation of indi-vidual fish at close range. Flexible sensors [10] can monitor individual behaviors, but ma-terial limitations prevent them from satisfying long-term monitoring needs, thus restrict-ing their application scope. Computer vision technology has been widely adopted in ab-normal behavior detection for aquaculture due to its noninvasive nature, real-time capa-bility, extensive monitoring capacity, and relatively low hardware costs [11]. However, behavior detection alone has apparent limitations, as it cannot track the behavioral changes in individuals across temporal sequences and struggles to distinguish individual differences within populations [12]. These challenges have prompted us to expand our research focus from static abnormal behavior detection to the dynamic field of abnormal behavior tracking, thereby achieving more comprehensive and continuous monitoring and analysis of fish abnormal behaviors [13–15].

Behavior tracking technology, as a natural extension of detection technology, integrates target detection, target association, and motion prediction to continuously track the identity and behavioral patterns of each fish in sequential video frames. Traditional be-havior tracking research has primarily focused on Correlation Filter (CF) technology, exemplified by filtering algorithms such as MOSSE [16], CSK [17], KCF [18], and BACF [19]. Although these methods are computationally efficient, their applications are limited. Their core Kalman filtering mechanism can only perform linear motion prediction based on coordinate information of target bounding boxes, lacking the capability to model and update target appearance features. This limitation results in drastically degraded tracking performance under conditions of rapid fish deformation or occlusion, failing to meet the requirements for group behavior analysis in practical aquaculture scenarios. Furthermore, due to the textural similarity among fish, feature engineering-based representation methods exhibit insufficient robustness in complex aquatic environments, leading to signifi-cant deviations in overall behavior analysis and tracking. This prevents effective macro-scopic description of fish behavioral states and makes it difficult to handle complex inter-active behaviors in multi-fish scenarios [20].

With the advancement of parallel computing capabilities and the development of deep learning technologies [21], efficient computation of complex models has become feasible, driving the rapid development of Multi-Object Tracking (MOT) technology. Multi-object tracking technology can provide more comprehensive, accurate, and real-time monitoring in fish behavioral anomaly detection, adapting to dynamic and complex aquaculture environments [22]. Deep learning-based target tracking algorithms are primarily classified into one-stage [23] and two-stage [24] approaches. Among these, one-stage target tracking algorithms integrate detection and appearance feature extraction branches into a single backbone, utilizing weight-sharing networks to simultaneously perform detection and feature extraction, then complete the tracking process through correlation filtering and data association, thereby enhancing the real-time performance of the algorithm. Examples include FairMOT and CenterTrack [25–27], which employ anchor-free detection strategies to collect embedding data required for tracking and transform it into classification information for assignment. However, due to insufficient ReID feature learning, these algorithms are prone to target loss and mismatching, resulting in lower tracking accuracy and difficulties in effectively handling complex morphological changes during fish swimming. Two-stage tracking methods are divided into two independent stages: target detection and target association. In the target detection stage, deep learning models such as RCNN [28] or YOLO [29] are employed to identify targets. In the association stage, cross-frame target recovery is accomplished based on appearance or motion cues of detected targets. Classic two-stage tracking algorithms include ByteTrack and DeepSORT. DeepSORT first utilizes target detection algorithms to identify objects and additionally trains convolutional neural networks to obtain appearance features of objects for feature matching. Du et al. [30] upgraded DeepSORT by adopting YOLOX-X and ResNeSt50 as the detection module and ReID module, respectively, effectively improving the model's tracking accuracy. ByteTrack simplifies the ReID dependency and achieves data association optimization through a dual-threshold detection box processing strategy, improving computational efficiency. However, the aforementioned two-stage tracking methods primarily focus on optimizing target detection and data association, and most existing re-search concentrates on scenarios such as human subjects and street vehicles, with insufficient consideration for the special requirements of fish behavior tracking in aquaculture. When fish exhibit abnormal physiological activities, they display nonlinear swimming patterns and severe mutual occlusion, and the similar appearance of fish bodies can easily cause similarity matching-based cross-frame detection methods to fail, resulting in frequent ID switching and inability to meet continuous tracking requirements, which con-strains the further application of underwater fish tracking technology.

This study proposes an improved DeepSORT strategy based on enhanced appearance feature matching. Experimental validation demonstrates that the proposed method achieves accurate tracking and identification of fish targets in complex underwater environments, with recognition performance significantly superior to other comparative models. In summary, our main contributions are as follows:

(1) ResNet is adopted as the backbone network to reconstruct the ReID feature extraction network for enhancing underwater fish multi-object tracking performance. Deformable Convolutional Networks v2 (DCNv2) [31] is embedded into residual blocks to achieve adaptive receptive field adjustment, thereby accurately capturing the nonlinear morphological changes of fish during the swimming motion.
(2) Triplet Loss function [32] is introduced as a training strategy to improve the robustness and discriminative capability of the ReID module among similar fish individuals.
(3) The Convolutional Block Attention Module (CBAM) [33] is incorporated to enhance feature learning of key fish body parts, effectively suppressing interference factors in complex underwater environments and further improving feature recognition accuracy.
(4) The effectiveness of the proposed method is comprehensively validated through ablation experiments and comparative analysis, providing a complete technical solution for underwater fish multi-object tracking.

The remainder of this paper is organized as follows: Section 2 briefly introduces the experimental methodology, data collection and processing procedures, and provides a detailed exposition of the proposed method; Section 3 conducts ablation experiments on real datasets and performs comparative analysis with existing state-of-the-art methods; Section 4 discusses and analyzes the experimental results, including the discussion of fish trajectory visualization results; Section 5 summarizes the work presented in this paper, analyzes the limitations of existing methods, and proposes future research directions.

2. Materials and Methods

2.1. Experimental Setup

The experiment was conducted at the laboratory of Sanya Marine Research Institute, Ocean University of China. The experimental subjects were leopard coral grouper with a body weight of (8.9 ± 2) g and body length of (9.6 ± 2) cm. The experiment utilized a recirculating aquaculture tank with a diameter of 1.5 m and depth of 0.8 m, equipped with a comprehensive recirculating water system, cooling water system, and real-time environmental monitoring system. Before the experiment, the coral trout were acclimated in the tank for 1–2 days to ensure no stress response. Natural seawater was used for cultivation, with initial water temperature maintained at (26 ± 1) °C and dissolved oxygen concentration controlled at (6.2 ± 0.2) mg/L. Fish were fed three times daily at regular intervals to meet their normal feeding requirements.

The low temperature stress experiment began at 26 °C with temperature reduction controlled by an intelligent temperature regulation system, decreasing by 1 °C every 2 h on average. Video recording was conducted throughout the entire process to document the transition of fish behavior from normal state to abnormal behaviors. The video acquisition system consisted of a digital video recorder and underwater cameras, as shown in Figure 1. A Boles UW-46S2-3LCTX10 camera (2 megapixels, 2 mm fixed focal length lens, focusing distance 0.6 m) made in Shenzhen, China, was mounted on a retractable bracket and positioned vertically in the center of the tank with horizontal placement and no elevation angle. The captured video resolution was 1920 × 2560 with a frame rate of 20 fps. The data analysis platform operated on a 64-bit Windows 10 system, equipped with an Intel(R) Xeon(R) CPU processor (from Intel Corporation, Santa Clara, CA, USA) and NVIDIA Quadro M4000 GPU (24GB VRAM, from NVIDIA Corporation, Santa Clara, CA, USA). The algorithm was based on the PyTorch 3.7 framework and developed using Python 3.9 programming language.

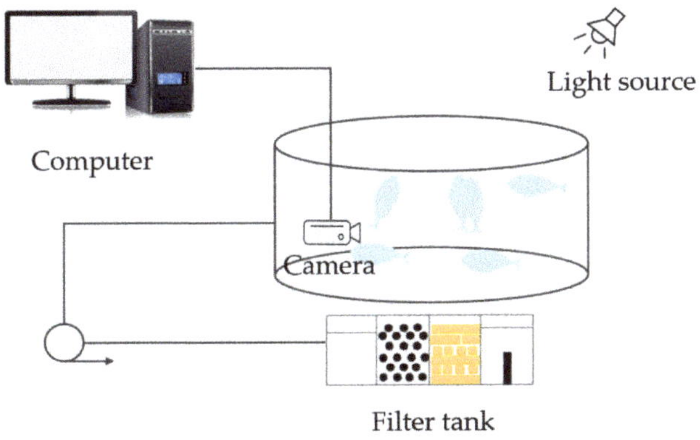

Figure 1. Image Acquisition System.

2.2. Dataset

During the data processing phase, we extracted image sequences from continuously recorded videos at a sampling rate of 15 frames per second to ensure continuous capturing of behavioral changes over time. Using the LabelImg tool, we performed fine-grained annotation on 3000 collected images of leopard coral grouper, focusing on marking typical patterns such as behavioral abnormalities and swimming anomalies, thereby constructing a complete abnormal behavior recognition dataset. To evaluate the performance of the multi-object tracking model, we additionally selected a representative group interaction behavior video segment containing 600 frames and performed frame-by-frame annotation following the MOTChallenge standard format, recording the bounding box coordinates and identity ID for each target.

This study employed the YOLOv12 model to construct an object detector for identifying surface-floating abnormal behavior in leopard coral grouper and trained it on the constructed abnormal behavior dataset. The dataset was randomly divided into training, validation, and test sets with a 7:2:1 ratio, consisting of 2100 training images, 600 validation images, and 300 test images, respectively. All datasets were validated through cross-validation to ensure balanced distribution, where the validation set was used for hyperparameter tuning, while the test set was strictly reserved for the final evaluation phase to ensure the reliability and accuracy of experimental results. To enhance model generalization and effectively mitigate overfitting, this study employs ImageNet pre-trained weights for model initialization and comprehensively applies multiple anti-overfitting techniques including data augmentation strategies and training optimization. During training, the Adam optimizer is utilized with β1 = 0.9 and β2 = 0.999, with an initial learning rate of 0.001 and a StepLR scheduler that reduces the learning rate by a factor of 0.1 every 30 epochs. The training epochs were set to 1000 to ensure sufficient model convergence, with a batch size of 16.

2.3. Research Methodology

2.3.1. Deformable Convolutional Networks

Fish targets exhibit morphological diversity and postural variation challenges. Due to the influence of water flow disturbances and swimming behaviors, individual fish specimens present irregular geometric deformations and postural differences at different temporal instances. Traditional standard convolution operations employ fixed rectangular receptive fields for feature extraction, making it difficult to effectively adapt to such dynamic morphological changes. This results in extracted feature representations lacking geometric adaptability, thereby affecting the accuracy of ReID.

To address the aforementioned geometric deformation adaptation problem, this study introduces DCNv2. As illustrated in Figure 2, DCNv2 enables convolution operations to flexibly adapt to geometric variations in images through dynamic adjustment of convolutional kernel sampling positions, thereby capturing more discriminative morphological features. Standard convolution operations employ fixed sampling patterns on input feature maps, with their output computation formula as follows:

$$y(b) = \sum_{k=1}^{K} \omega_k \cdot A(b + b_k), \tag{1}$$

In this formula, b_k represents the k-th predefined fixed offset relative to position b in the convolutional kernel, ω_k denotes the convolutional weight at the k-th position, and K is the kernel size.

DCNv2 introduces a learnable offset and modulation factor mechanism based on ordinary convolution, with the output equation:

$$y(b) = \sum_{k=1}^{K} \omega_k \cdot X(b + b_k + \Delta b_k) \cdot \Delta m_k, \tag{2}$$

where Δb_k is the learnable offset of the k-th sampling point, and Δm_k is the modulation factor, which is generated by independent branching via the sigmoid function and is used to control the feature weights of each sampling point.

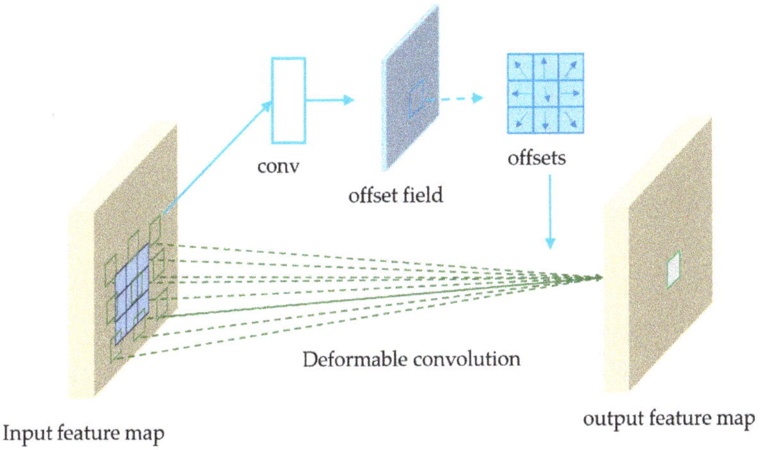

Figure 2. Deformable Convolutional Networks v2.

2.3.2. Improved Residual Structure

As illustrated in Figure 3, residual blocks effectively mitigate the vanishing gradient problem in deep neural networks through the introduction of skip connections, enabling direct information propagation from shallow to deep layers and ensuring that networks maintain effective training performance while increasing depth. To address the flexible deformation characteristics of moving fish targets in underwater environments and their dynamic feature matching requirements, this paper presents improvements to the traditional residual structure. We integrate DCNv2 into residual blocks, replacing the original standard convolutional layers to enhance the network's capacity for modeling geometric transformations. Unlike standard convolution with fixed receptive fields, DCNv2 can learn additional spatial offset parameters and dynamically adjust the sampling positions of convolutional kernels, making it more suitable for capturing non-rigid deformation features of fish.

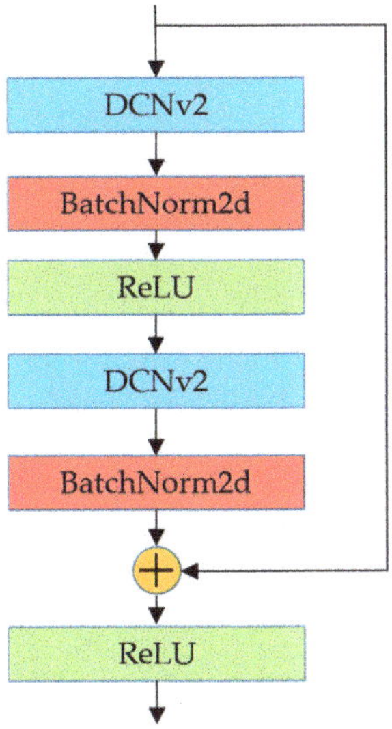

Figure 3. The residual block structure integrated with deformable convolution.

2.3.3. Triplet Loss Function

During fish target tracking, frequent ID switching occurs due to the high similarity in fish appearance. Triplet Loss optimizes feature space distribution by constructing triplet-based loss, as shown in Figure 4. The triplet consists of three components: the anchor feature vector (Anchor), a positive sample feature vector from the same category as the Anchor (Positive), and a negative sample feature vector from a different category than the Anchor (Negative). Through this approach, the network learns to generate a discriminative embedding space where each input is mapped to a unique, distinguishable feature vector. By training the network to maximize the distance between feature vectors of similar targets and minimize the distance between dissimilar targets, ID switching caused by similar target appearances is effectively reduced. The formula for Triplet Loss is as follows:

$$L_{triplet} = \max(d(a,p) - d(a,n) + \alpha, 0), \tag{3}$$

where d(a,p) represents the distance between the Anchor and Positive samples, d(a,n) represents the distance between the Anchor and Negative samples, and α is a hyperparameter used to control the distance requirement for negative samples. Triplet Loss prevents the embedding vectors of positive and negative samples from being trained too closely together by setting a margin constant hyperparameter, which is set to 0.5 in this study.

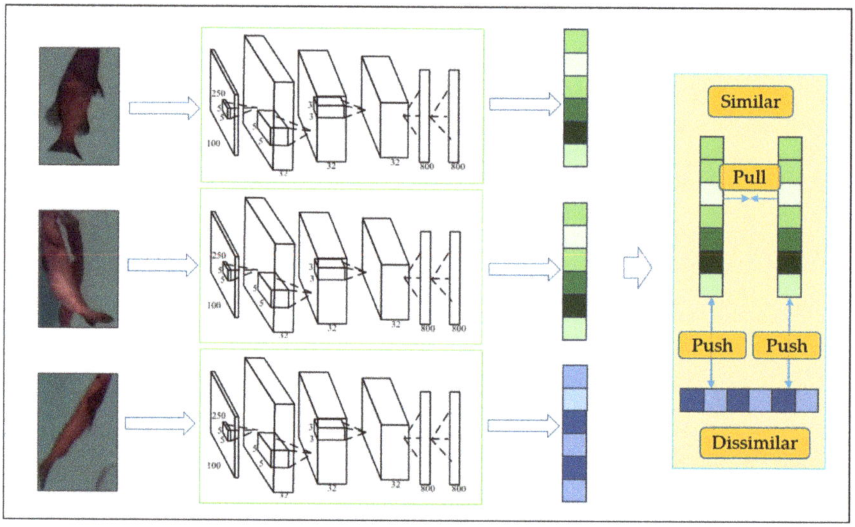

Figure 4. Triplet Loss Training Framework.

2.3.4. Enhanced ReID Feature Extraction Network

The DeepSORT algorithm introduces a ReID model based on the SORT framework, enhancing the matching effectiveness of the Hungarian algorithm by combining appearance information with motion information [34]. The original ReID network is primarily designed for pedestrian detection scenarios, with network architecture and input image dimensions (64 × 128) optimized for human characteristics [35]. However, when applied to underwater fish target detection, significant differences exist compared to human and vehicle tracking due to the nonlinear motion, variable morphology, and highly similar appearance characteristics of underwater fish. Relying solely on color and texture recognition yields poor results, leading to suboptimal performance of traditional ReID networks in underwater fish target association. To address these issues, this paper proposes an improved ReID feature extraction network strategy based on appearance feature optimization, aimed at enhancing feature matching efficiency and constraining the excessive growth of underwater target IDs. Figure 5 illustrates the improved feature extraction network architecture.

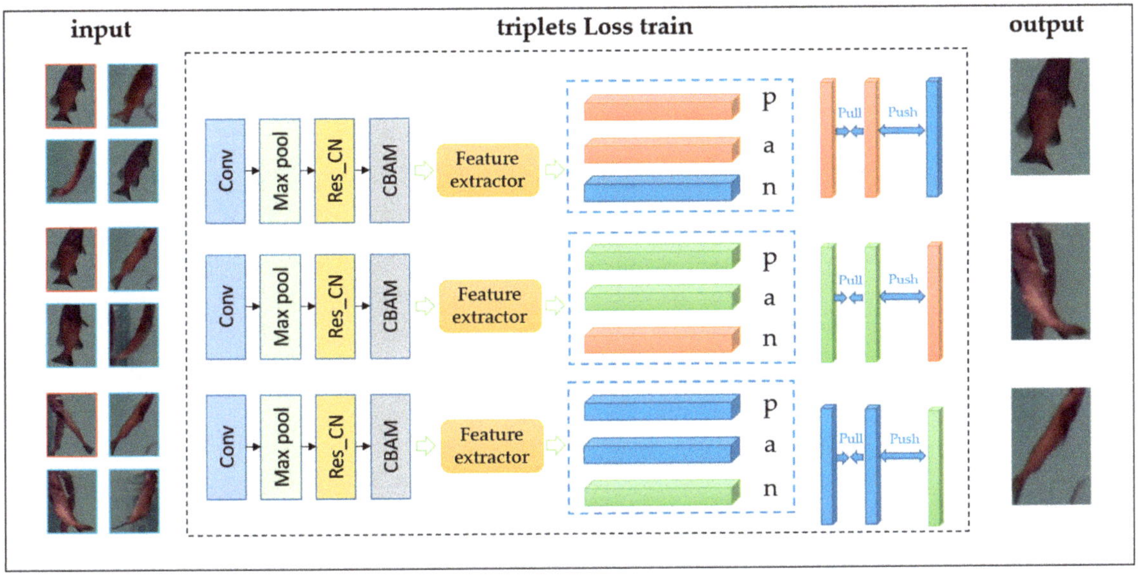

Figure 5. Enhanced Feature Extraction Network Architecture for ReID.

Residual networks effectively mitigate the gradient vanishing problem in deep networks through skip connections. In this study, DCNv2 is introduced to replace the original structure in the first 3×3 convolutional layer of the third residual block within the feature extraction network. The incorporation of DCNv2 further enhances the network's adaptability to complex spatial structures, enabling the model to more flexibly capture structural information in images, which is particularly suitable for processing underwater fish targets with non-rigid deformation characteristics.

The CBAM is introduced after the improved residual block. The CBAM employs a cascaded design of channel attention and spatial attention, capable of adaptively recalibrating feature responses. Through the channel attention mechanism, the network focuses on the most important feature channels for the current task, effectively suppressing redundant information interference. The spatial attention mechanism concentrates on the most discriminative spatial locations in feature maps, highlighting key regional features of targets.

This synergistic effect of dual attention and deformable convolution enables the network to extract fish key features more precisely, effectively filtering background noise in complex underwater environments, and providing high-quality feature representations for subsequent triplet loss optimization. To further address the high inter-similarity problem among underwater targets, this study employs a triplet loss function to perform deep optimization on the outputs of the aforementioned feature extraction module. The triplet loss forms an end-to-end optimization framework with the DCNv2-CBAM feature extraction module, where the feature extraction module provides rich and discriminative feature representations, while the triplet loss further refines the discriminative boundaries in the feature space by simultaneously optimizing positive sample pair similarity and negative sample pair dissimilarity. This training mechanism fully exploits the refined features extracted by the DCNv2-CBAM, controlling the relative distances between positive and negative sample pairs in the embedding space through the margin parameter, thereby preventing excessive clustering of feature vectors that would compromise discriminative capability. The introduction of triplet loss enables the entire network to learn more robust and highly discriminative deep feature representations, significantly improving the accuracy and reliability of ReID.

3. Results

3.1. Evaluation Metrics

To evaluate the results, the model is applied to the CLEAR metrics used on the MOT Challenge to assess model performance [36,37]. The following are some typical evaluation metrics: Multiple Object Tracking Accuracy (MOTA) [38] is one of the most important metrics in multi-object tracking. Identity switches (IDsw) are the only tracking related component in MOTA; thus, it only considers the accuracy of instance matching. The calculation formula is as follows:

$$\text{MOTA} = 1 - \frac{\sum_t (FN_t + FP_t + IDS_t)}{\sum_t GT_t}, \tag{4}$$

In the formula, FN_t, FP_t, IDS_t, and GT_t represent the number of false negatives, false positives, identity switches, and ground truth targets in frame t, respectively.

The IDF1 [39] identification score (IDF1) comprehensively considers identification precision (IDP) and identification recall (IDR). In IDF1, the involved true positives (*TP*), false positives (*FP*), and false negatives (*FN*) all take into account the ID information, making this metric more sensitive to the accuracy of ID information. The calculation formula is as follows:

$$\text{IDF1} = \frac{TP}{TP + 0.5FP + 0.5FN}, \tag{5}$$

Multiple Object Tracking Precision (MOTP) represents the localization accuracy of the detector, and its calculation formula is as follows:

$$\text{MOTP} = \frac{\sum_{t,i} d_{t,i}}{C_t}, \tag{6}$$

where C_t denotes the number of correctly matched objects at frame t, and $d_{t,i}$ represents the bounding box overlap between target i and its assigned ground truth object. A smaller MOTP indicates lower tracking localization error.

3.2. Ablation Experiment

The ablation experiments in this section demonstrate that the Res-ReID network effectively addresses the gradient problem inherent in the original ReID feature extraction network. The Triplet Loss enhances fish individual identification capability, DCN improves adaptability to deformed targets, and CBAM strengthens key feature extraction. Through the synergistic effects of multiple modules, the system achieves a 4.15% improvement in MOTA metrics and an 18.55% improvement in IDF1 metrics compared to the original DeepSORT, thereby validating the effectiveness of the proposed improvement scheme. Figure 6 illustrates the tracking performance of the system in practical applications.

In the original ReID network, the individual incorporation of Triplet Loss resulted in an improvement of IDF1 from the baseline value of 53.73% to 61.73%, while simultaneously enhancing IDP from 51.98% to 62.73% and IDR from 54.36% to 61.26%. Under the Res-ReID network architecture, the introduction of Triplet Loss improved IDF1 from 53.73% to 62.90%. When Triplet Loss was combined with DCN, the effects were more pronounced.

In the original network structure, IDF1 improved from 61.73% to 62.47%, with IDP and IDR increasing from 62.71% and 61.26% to 63.90% and 61.73%, respectively. Under the improved ReID network architecture, the combined application of these two techniques enhanced IDP and IDR from 63.24% and 62.50% to 64.67% and 62.91%, respectively.

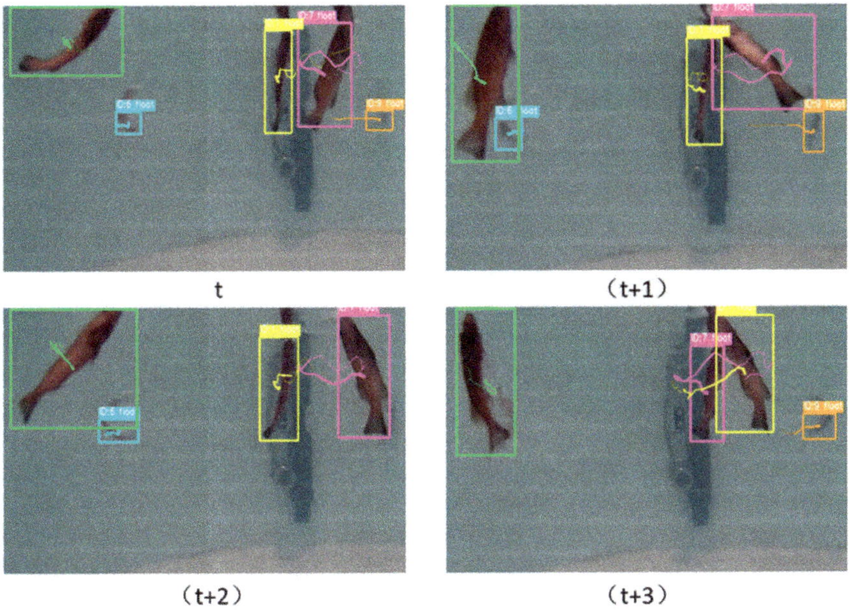

Figure 6. Multi-object tracking trajectories of the improved model across consecutive frames t, t + 1, t + 2, and t + 3, with different colors representing different tracked objects.

When DCN was applied independently, the MOTA of the original network improved from 64.26% to 64.98%, while in the Res-ReID network, MOTA increased from 64.57% to 65.10%. When DCN was combined with Triplet Loss, a favorable synergistic effect was observed. In the original network structure, this combination enhanced MOTA from 64.26% to 64.96% and improved IDF1 from 53.73% to 62.47%. In the Res-ReID network structure, compared to individual application, MOTA increased from 65.10% to 66.56%, and IDF1 also improved from 60.00% to 62.85%.

Following the introduction of CBAM into the baseline ReID model, as shown in Table 1, MOTA, IDF1, and IDP increased by 0.56%, 1.35%, and 1.44%, respectively. In the Res-ReID network structure, CBAM similarly enhanced all performance metrics, with MOTA, IDF1, and IDR improving by 0.63%, 0.58%, and 0.92%, respectively. These results validate the effectiveness of the CBAM attention mechanism in multi-object tracking tasks.

Table 1. Comparison of Ablation Experiment Results.

Algorithm	MOTA	IDF1	Recall	IDP	IDR
ReID	64.26%	53.73%	80.70%	51.98%	54.36%
ReID + Triplets Loss	63.95%	61.73%	80.70%	62.73%	61.26%
ReID + DCN	64.98%	54.93%	81.20%	54.65%	57.30%
ReID + CBAM	64.79%	54.56%	80.68%	53.36%	54.42%
ReID + Triplets Loss + DCN	64.96%	62.47%	81.00%	63.90%	61.73%
ReID + Triplets Loss + DCN + CBAM	65.37%	62.83%	81.20%	63.29%	62.30%

Table 1. *Cont.*

Algorithm	MOTA	IDF1	Recall	IDP	IDR
Res-ReID + Triplets Loss	64.57%	62.90%	80.90%	63.24%	62.50%
Res-ReID + DCN	65.10%	60.00%	80.00%	60.97%	61.32%
Res-ReID + CBAM	65.98%	60.24%	81.41%	61.81%	59.33%
Res-ReID + Triplets Loss + DCN	66.56%	62.85%	81.24%	64.67%	62.91%
Res-ReID + Triplets Loss + DCN + CBAM	66.93%	63.70%	81.40%	65.60%	62.43%

3.3. Comparison with Other Models

To systematically evaluate the effectiveness of the proposed method, this study selected four classical algorithms—SORT [40], DeepSORT [41], ByteTrack [42], and FairMOT [43]—for comparative analysis, covering both one-stage and two-stage tracking. All methods were tested on a unified dataset to ensure fair evaluation of their detection and ReID performance. As shown in Table 2, ByteTrack, using the same detector as our model, achieved a MOTA of 64.56%, but with an IDF1 of only 36.52%. The proposed method demonstrates improvements in MOTA of 72.64%, 3.67%, and 5.07% compared to SORT, ByteTrack, and FairMOT, respectively. For IDF1, our method shows improvements of 79.94%, 74.48%, and 37.52% compared to SORT, ByteTrack, and FairMOT, respectively. Both IDP and IDR metrics exhibit improvements of over 35% across all comparisons.

Table 2. Comparison of the Method in This Study with Other Methods. Note: In the table, ↑ indicates that higher values are better, and ↓ indicates that lower values are better.

Method	MOTA ↑	IDF1 ↑	MOT ↓	IDP ↑	IDR ↑
Sort	38.78%	35.40%	29.36%	35.97%	34.31%
Bytetrack	64.56%	36.52%	25.84%	38.70%	34.48%
FairMOT	63.70%	46.32	27.10%	46.53%	45.06%
Proposed Method	66.93%	63.70%	25.91%	65.60%	62.43%

4. Discussion

As demonstrated in Table 1, the incorporation of Triplet Loss resulted in improvements across IDF1, IDP, and IDR metrics, with a particularly notable enhancement of 17.04% in IDF1 under the Res-ReID network architecture. These findings indicate that Triplet Loss enhances the model's identity discrimination capability. However, while identity recognition-related metrics showed improvement, the MOTA metric remained unchanged. This phenomenon can be attributed to the fact that MOTA evaluates the overall accuracy of multi-object tracking, encompassing factors such as missed detections, false positives, and identity switches, whereas Triplet Loss may exhibit excessive focus on identity features in scenarios involving frequent interactions, potentially neglecting localization accuracy or detection completeness. As illustrated in Figure 7, when applied to the Res-ReID network, the advantages of Triplet Loss are further amplified, resulting in comprehensive tracking performance enhancement. As shown in Table 1, when DCN is applied individually to both the original network and the Res-ReID network, improvements in MOTA are observed. This enhancement stems from DCN's capability to dynamically adjust the receptive field according to target deformation, thereby effectively strengthening the model's ability to handle flexible motion in fish targets. However, when DCN is employed independently, the improvement in identity recognition-related metrics such as IDF1, IDP, and IDR is not significant. As depicted in Figure 7, although DCN enhances the model's feature extraction capability, certain limitations persist in terms of feature discrimination and maintaining identity consistency. When Triplet Loss is combined with DCN, both the original ReID model and Res-ReID demonstrate more substantial improvements, particularly in IDP and

IDR metrics. This improvement occurs because DCN enhances the model's adaptability to target deformation, while Triplet Loss provides superior feature discrimination capability, creating a favorable complementary effect between these two components. The CBAM attention mechanism filters the most discriminative features to provide more precise input for Triplet Loss, while simultaneously collaborating with DCN to handle target deformation and complex motion trajectories.

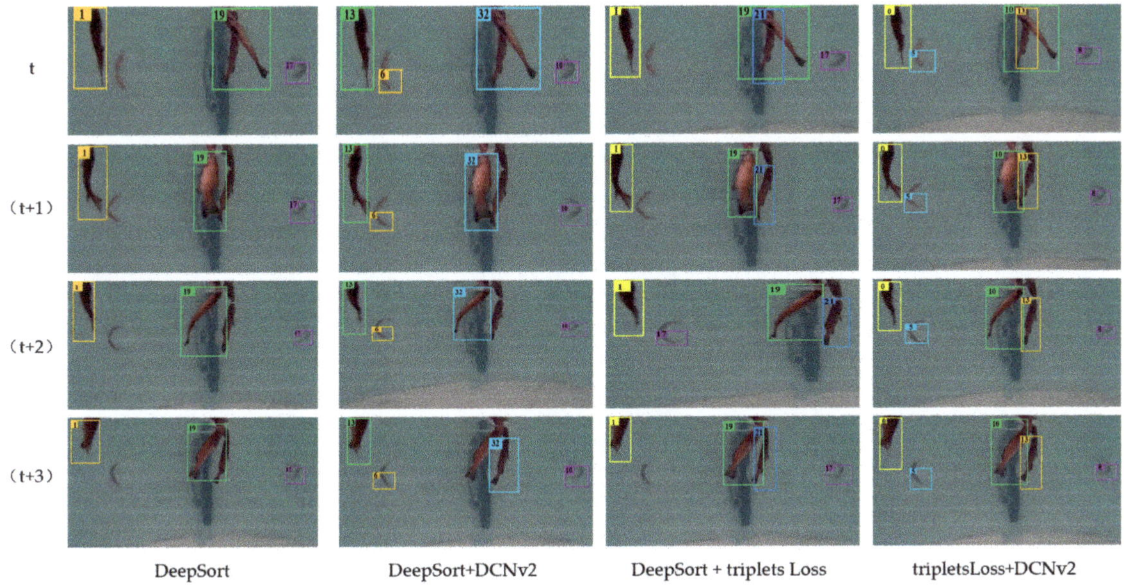

Figure 7. Visualization of tracking results in ablation study. From left to right: DeepSort, DeepSort + DCNv2, DeepSort + Triplet Loss, and Triplet Loss + DCNv2. Each method shows tracking results across four consecutive frames (t, t + 1, t + 2, t + 3) with unique ID numbers for each fish, demonstrating tracking continuity.

As demonstrated in Table 2, ByteTrack achieved a MOTA of 64.56% using the same detector as our model but obtained an IDF1 of only 36.52%. This indicates that while the algorithm exhibits satisfactory tracking accuracy, the absence of a dedicated ReID module causes the association mechanism to fail when handling similar targets, resulting in relatively low IDF1 performance. Consequently, ByteTrack is more suitable for relatively simple multi-object tracking tasks. FairMOT employs a joint training paradigm attempting to balance detection and tracking tasks; however, its model lacks a specialized tracking branch and relies solely on Kalman filtering for motion modeling. In the specific task of fish tracking, it struggles to effectively distinguish between fish targets with similar appearances. Figure 8 illustrates the fish movement trajectories across different multi-object tracking models. In the experimental scenario, the fish 1 and fish 4 exhibit relatively simple lateral swimming patterns with minimal occlusion interference. The fish 2 and fish 3 demonstrate high-frequency interaction states, continuously performing rotational movements with frequent mutual occlusion. This complex interactive behavior results in highly nonlinear and irregular motion trajectory characteristics. Through comparative analysis of the tracking trajectory patterns across different algorithms, it is evident that the traditional SORT model suffers from insufficient detection accuracy and target ReID capability in complex underwater environments, making it difficult to maintain target identity continuity. Although FairMOT and ByteTrack show improvements in overall performance compared to SORT, they still exhibit technical limitations when confronted with the complex scenarios in our experiments. Both methods demonstrate frequent ID switching phenomena, with trajectory fragmentation issues remaining prominent. In contrast, the improved algorithm proposed in this paper generates more continuous and complete motion trajectories for the fish 1 and fish 2 tracking, with significantly reduced ID switching frequency compared to baseline methods, demonstrating superior stability and robustness. However, for the fish 3 under conditions of prolonged, high-frequency mutual occlusion, the performance remains suboptimal.

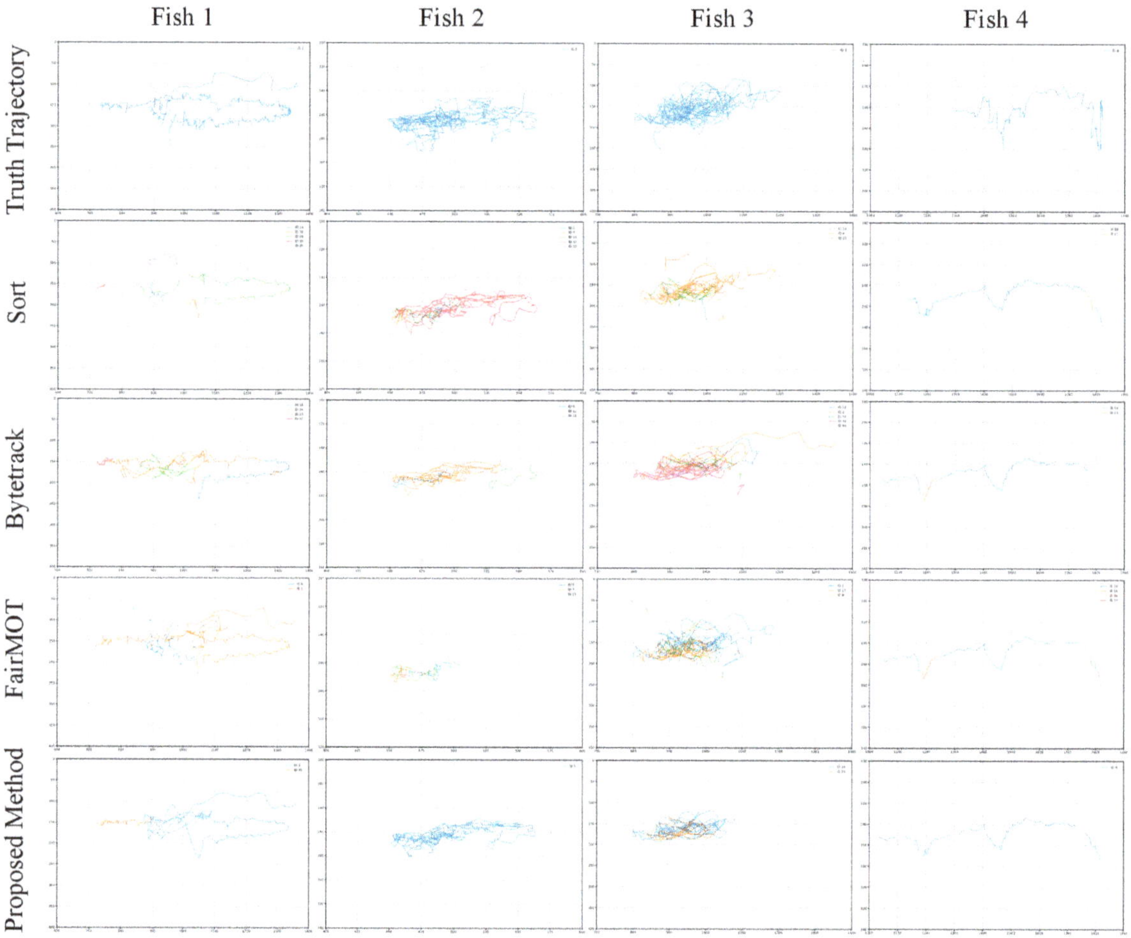

Figure 8. Trajectory Comparison of Multiple Models. Note: The figure displays tracking results of five methods arranged by rows: Truth Trajectory, Sort, ByteTrack, FairMOT, and the Proposed Method; columns correspond to four tracking targets (Fish 1–4). In each subplot, continuous same-colored line segments represent stable identity tracking, while color transition points indicate identity switch errors.

This experiment was conducted under relatively ideal experimental conditions. While it successfully validates the fundamental effectiveness of the algorithm, it fails to adequately simulate the complex collective behavioral characteristics observed in high-density aquaculture environments. In real aquaculture scenarios, highly dense population structures and frequent complex interactive behaviors among individual fish result in inter-target occlusion levels and durations that far exceed those encountered in experimental settings. The model's capability to handle complex occlusion scenarios requires further enhancement. As illustrated in Figure 8, the tracking performance for the fish 3 remains suboptimal under conditions of prolonged, high-frequency occlusion. The severe loss of target feature information presents a significant challenge: when fish targets are completely occluded by other individuals for extended periods, available visual features are substantially reduced, making it difficult for the ReID network to extract sufficient discriminative features. Therefore, how to further improve algorithm performance under extreme occlusion conditions represents a critical technical challenge that must be addressed in future research endeavors.

5. Conclusions

This study addresses the challenges of similar texture patterns and frequent occlusion-induced ID switching in high-density aquaculture fish tracking by proposing an improved DeepSORT tracking model that integrates deformable convolution and CBAM attention modules with triplet loss training. The proposed method effectively resolves occlusion and trajectory fragmentation issues in underwater fish tracking through occlusion detection and target ReID mechanisms. The effectiveness of this approach is validated through ablation experiments. Performance comparisons with other state-of-the-art models demonstrate that the optimized DeepSORT framework exhibits superior performance. However, tracking performance in scenarios involving frequent fish swarm interactions remains to be improved. How to further enhance the tracking stability and accuracy of the algorithm under extreme occlusion conditions represents a core technical challenge that requires focused breakthrough efforts in future research endeavors.

Author Contributions: Writing—original draft, S.L. (Shengnan Liu); data curation, J.Z. and S.L. (Shengnan Liu); formal analysis, H.Z.; software, S.L. (Shengnan Liu) and J.Z.; investigation, formal analysis, C.Q. and S.L. (Shijing Liu). All authors have read and agreed to the published version of the manuscript.

Funding: Supported by Hainan Seed Industry Laboratory B23H10004; Central Public-interest Scientific Institution Basal Research Fund, CAFS (NO. 2024XT0901); Central Public-interest Scientific Institution Basal Research Fund, ECSFRCAFS (NO. 2025YJ02).

Institutional Review Board Statement: This study complied with the regulations and guidelines established by the Animal Care and Use Committee of Fishery Machinery and Instrument Research Institute, Chinese Academy of Fishery Sciences (ethics committee approval code: FMIRI-AWE-2024-002, approval date: 7 August 2024).

Data Availability Statement: The datasets presented in this article are not readily available because the data are part of an ongoing study. Requests to access the datasets should be directed to liushijing@fmiri.ac.cn.

Conflicts of Interest: The authors declare no conflicts of interest.

References

1. FAO. *The State of World Fisheries and Aquaculture 2024*; Blue Transformation in Action; FAO: Rome, Italy, 2024.
2. Ruan, W.; Wang, Y.; Ji, W.W.; Han, B.; Fang, H. Progress of sustainable development and management of aquaculture. *Fish. Inf. Strat.* **2013**, *28*, 267–272.
3. Xu, H.; Jiang, T. Development strategy of offshore aquaculture engineering in China. *Fish. Mod.* **2012**, *39*, 1–7.
4. Cui, M.; Jin, J.; Huang, W. Discussion about system construction and general technology of aquaculture platform. *Fish. Mod.* **2019**, *46*, 61–66.
5. Mwaffo, V.; Vernerey, F. Analysis of group of fish response to startle reaction. *J. Nonlinear Sci.* **2022**, *32*, 96. [CrossRef]

6. Barreiros, M.D.O.; Dantas, D.D.O.; Silva, L.C.D.O.; Ribeiro, S.; Barros, A.K. Zebrafish tracking using YOLOv2 and Kalman filter. *Sci. Rep.* **2021**, *11*, 3219. [CrossRef]
7. Huang, Y.X.; Bao, X.T.; Xu, H. Research progress of fishery equipment science and technology in China. *Fish. Mod.* **2023**, *50*, 1–11.
8. Li, D.L.; Wang, G.X.; Du, L.; Zheng, Y.; Wang, Z. Recent advances in intelligent recognition methods for fish stress behavior. *Aquac. Eng.* **2022**, *96*, 102222. [CrossRef]
9. Liu, T.C.; Liu, J.; Wang, J.; Xu, J.; Sun, G. Optimization of the intelligent sensing model for environmental information in aquaculture waters based on the 5G smart sensor network. *J. Sens.* **2022**, *2022*, 6409046. [CrossRef]
10. Makaras, T.; Razumienė, J.; Gurevičienė, V.; Šakinytė, I.; Stankevičiūtė, M.; Kazlauskienė, N. A new approach of stress evaluation in fish using β-D-Glucose measurement in fish holding-water. *Ecol. Indic.* **2020**, *109*, 105829. [CrossRef]
11. Davidson, J.; Good, C.; Welsh, C.; Summerfelt, S.T. Abnormal swimming behavior and increased deformities in rainbow trout Oncorhynchus mykiss cultured in low exchange water recirculating aquaculture systems. *Aquac. Eng.* **2011**, *45*, 109–117. [CrossRef]
12. Leon, F.; Gavrilescu, M. A review of tracking and trajectory prediction methods for autonomous driving. *Mathematics* **2021**, *9*, 660. [CrossRef]
13. Shreesha, S.; Manohara, P.M.M.; Verma, U.; Pai, R.M. Computer vision based fish tracking and behaviour detection system. In Proceedings of the 2020 IEEE International Conference on Distributed Computing, VLSI, Electrical Circuits and Robotics (DISCOVER), Udupi, India, 30–31 October 2020; pp. 252–257. [CrossRef]
14. Kumar, A.; Walia, G.S.; Sharma, K. Recent trends in multi-cue based visual tracking: A review. *Expert Syst. Appl.* **2020**, *162*, 113711. [CrossRef]
15. Sun, S.; Akhtar, N.; Song, H.; Mian, A.; Shah, M. Deep affinity network for multiple object tracking. *IEEE Trans. Pattern Anal. Mach. Intell.* **2021**, *43*, 104–119. [CrossRef]
16. Bolme, D.S.; Beveridge, J.R.; Draper, B.A.; Lui, Y.M. Visual object tracking using adaptive correlation filters. In Proceedings of the 2010 IEEE Computer Society Conference on Computer Vision and Pattern Recognition, San Francisco, CA, USA, 13–18 June 2010; pp. 2544–2550.
17. Rahmat, B.; Waluyo, M.; Rachmanto, T.A.; Afandi, M.I.; Widyantara, H.; Harianto. Video-based Tancho Koi Fish tracking system using CSK, DFT, and LOT. *J. Phys. Conf. Ser.* **2020**, *1569*, 022036. [CrossRef]
18. Tang, M.; Yu, B.; Zhang, F.; Wang, J. High-speed tracking with multi-kernel correlation filters. In Proceedings of the 2018 IEEE/CVF Conference on Computer Vision and Pattern Recognition, Salt Lake City, UT, USA, 18–23 June 2018; pp. 4874–4883. [CrossRef]
19. Kiani Galoogahi, H.; Fagg, A.; Lucey, S. Learning background-aware correlation filters for visual tracking. In Proceedings of the 2017 IEEE International Conference on Computer Vision (ICCV), Venice, Italy, 22–29 October 2017; pp. 1135–1143. [CrossRef]
20. Luo, W.; Xing, J.; Milan, A.; Zhang, X.; Liu, W.; Zhao, X.; Kim, T.-K. Multiple object tracking: A literature review. *Artif. Intell.* **2021**, *293*, 103448. [CrossRef]
21. Li, D.L.; Liu, C. Recent advances and future outlook for artificial intelligence in aquaculture. *Smart Agric.* **2020**, *2*, 1–20.
22. Stauffer, C.; Grimson, W.E.L. Adaptive background mixture models for real-time tracking. In Proceedings of the 1999 IEEE Computer Society Conference on Computer Vision and Pattern Recognition (Cat. No PR00149), Fort Collins, CO, USA, 23–25 June 1999; Volume 2, pp. 246–252. [CrossRef]
23. Emami, P.; Pardalos, P.M.; Elefteriadou, L.; Ranka, S. Machine learning methods for data association in multi-object tracking. *ACM Comput. Surv.* **2020**, *53*, 1–34. [CrossRef]
24. Wang, Z.; Zheng, L.; Liu, Y.; Wang, S. Towards real-time multi-object tracking. In Proceedings of the European Conference on Computer Vision, Glasgow, UK, 23–28 August 2020; Springer International Publishing: Cham, Switzerland, 2020; pp. 107–122.
25. Zhang, Y.; Wang, C.; Wang, X.; Zeng, W.; Liu, W. Fairmot: On the fairness of detection and re-identification in multiple object tracking. *Int. J. Comput. Vis.* **2021**, *129*, 3069–3087. [CrossRef]
26. Zhang, Y.; Sun, P.; Jiang, Y.; Yu, D.; Weng, F.; Yuan, Z.; Luo, P.; Liu, W.; Wang, X. Bytetrack: Multi-object tracking by associating every detection box. In Proceedings of the European Conference on Computer Vision, Tel Aviv, Israel, 23–27 October 2022; Springer Nature: Cham, Switzerland, 2022; pp. 1–21.
27. Zhou, X.; Koltun, V.; Krähenbühl, P. Tracking objects as points. In Proceedings of the European Conference on Computer Vision, Glasgow, UK, 23–28 August 2020; Springer International Publishing: Cham, Switzerland, 2020; pp. 474–490.
28. Ren, S.; He, K.; Girshick, R.; Sun, J. Faster r-cnn: Towards real-time object detection with region proposal networks. *Adv. Neural Inf. Process. Syst.* **2015**, *28*, 91–99. [CrossRef]
29. Bochkovskiy, A.; Wang, C.-Y.; Liao, H.-Y.M. Yolov4: Optimal speed and accuracy of object detection. *arXiv* **2004**, arXiv:2004.10934. [CrossRef]
30. Du, Y.; Zhao, Z.; Song, Y.; Zhao, Y.; Su, F.; Gong, T.; Meng, H. Strongsort: Make deepsort great again. *J. IEEE Trans. Multimed.* **2023**, *25*, 8725–8737. [CrossRef]

31. Zhu, X.; Hu, H.; Lin, S.; Dai, J. Deformable convnets v2: More deformable, better results. In Proceedings of the IEEE/CVF Conference on Computer Vision and Pattern Recognition, Long Beach, CA, USA, 15–20 June 2019; pp. 9308–9316.
32. Hermans, A.; Beyer, L.; Leibe, B. In defense of the triplet loss for person re-identification. *arXiv* **2017**, arXiv:1703.07737.
33. Bayar, E.; Aker, C. When to extract ReID features: A selective approach for improved multiple object tracking. *arXiv* **2024**, arXiv:2409.06617.
34. Zhang, X.; Hao, X.; Liu, S.; Wang, J.; Xu, J.; Hu, J. Multi-target tracking of surveillance video with differential YOLO and DeepSort. In Proceedings of the Eleventh International Conference on Digital Image Processing (ICDIP 2019), Guangzhou, China, 10–13 May 2019; Volume 11179, pp. 701–710.
35. Woo, S.; Park, J.; Lee, J.Y.; Kweon, I.S. Cbam: Convolutional block attention module. In Proceedings of the European Conference on Computer Vision (ECCV), Munich, Germany, 8–14 September 2018; pp. 3–19.
36. Dendorfer, P.; Rezatofighi, H.; Milan, A.; Shi, J.; Cremers, D.; Reid, I.; Roth, S.; Schindler, K.; Leal-Taixé, L. Mot20: A benchmark for multi object tracking in crowded scenes. *arXiv* **2020**. [CrossRef]
37. Bernardin, K.; Stiefelhagen, R. Evaluating multiple object tracking performance: The clear mot metrics. *EURASIP J. Image Video Process.* **2008**, *2008*, 1–10. [CrossRef]
38. Wu, B.; Nevatia, R. Tracking of multiple, partially occluded humans based on static body part detection. In Proceedings of the 2006 IEEE Computer Society Conference on Computer Vision and Pattern Recognition (CVPR'06), New York, NY, USA, 17–22 June 2006; Volume 1, pp. 951–958. [CrossRef]
39. Shin, L.J.; Lo, J.C.; Chen, G.H.; Callis, J.; Fu, H.; Yeh, K.-C. IRT1 degradation factor1, a ring E3 ubiquitin ligase, regulates the degradation of iron-regulated transporter1 in Arabidopsis. *Plant Cell.* **2013**, *25*, 3039–3051. [CrossRef] [PubMed]
40. Bewley, A.; Ge, Z.; Ott, L.; Ramos, F.; Upcroft, B. Simple online and realtime tracking. In Proceedings of the 2016 IEEE International Conference on Image Processing (ICIP), Phoenix, AZ, USA, 25–28 September 2016; pp. 3464–3468. [CrossRef]
41. Wojke, N.; Bewley, A.; Paulus, D. Simple online and realtime tracking with a deep association metric. In Proceedings of the 2017 IEEE International Conference on Image Processing (ICIP), Beijing, China, 17–20 September 2017; pp. 3645–3649. [CrossRef]
42. Aharon, N.; Orfaig, R.; Bobrovsky, B.Z. BoT-SORT: Robust associations multi-pedestrian tracking. *arXiv* **2022**. [CrossRef]
43. Chen, Y.; Chen, Z.; Zhang, Z.; Bian, S. AdapTrack: An adaptive FairMOT tracking method applicable to marine ship targets. *J. AI Commun.* **2023**, *36*, 127–145. [CrossRef]

Journal of
Marine Science and Engineering

MDPI

Article

The Hydrodynamic Performance of a Vertical-Axis Hydro Turbine with an Airfoil Designed Based on the Outline of a Sailfish

Aiping Wu [1,2,*,†], Shiming Wang [1,†] and Chenglin Ding [3]

1 College of Engineering Science and Technology, Shanghai Ocean University, Shanghai 201308, China; smwang@shou.edu.cn
2 Yangtze Aotecar New Energy Technology (Wuhan) Co., Ltd., Wuhan 430223, China
3 School of Artificial Intelligence, Shanghai Normal University Tianhua College, Shanghai 201815, China; cldingme@163.com
* Correspondence: jsdywap@hotmail.com; Tel.: +86-151-2104-8945
† These authors contributed equally to this work.

Abstract

This study investigates an aerodynamic optimization framework inspired by marine biological morphology, utilizing the sailfish profile as a basis for airfoil configuration. Through Latin hypercube experimental design combined with optimization algorithms, four key geometric variables governing the airfoil's hydrodynamic characteristics were systematically analyzed. Parametric studies revealed that pivotal factors including installation angle significantly influenced the fluid dynamic performance metrics of lift generation and pressure drag. Response surface methodology was employed to establish predictive models for these critical performance indicators, effectively reducing computational resource consumption and experimental validation costs. The refined bio-inspired configuration demonstrated multi-objective performance improvements compared to the baseline configuration, validating the computational framework's effectiveness for hydrodynamic profile optimization studies. Furthermore, a coaxial dual-rotor vertical axis turbine configuration was developed, integrating centrifugal and axial-flow energy conversion mechanisms through a shared drivetrain system. The centrifugal rotor component harnessed tidal current kinetic energy while the axial-flow rotor module captured wave-induced potential energy. Transient numerical simulations employing dynamic mesh techniques and user-defined functions within the Fluent environment were conducted to analyze rotor interactions. Results indicated the centrifugal subsystem demonstrated peak hydrodynamic efficiency at a 25° installation angle, whereas the axial-flow module achieves optimal performance at 35° blade orientation. Parametric optimization revealed maximum energy extraction efficiency for the centrifugal rotor occurs at $\lambda = 1.25$ tip-speed ratio under $Re = 1.3 \times 10^5$ flow conditions, while the axial-flow counterpart attained optimal performance at $\lambda = 1.5$ with $Re = 5.5 \times 10^4$. This synergistic configuration demonstrated complementary operational characteristics under marine energy conversion scenarios.

Keywords: sailfish; vertical axis water turbine; sharp speed ratio; optimized Latin hypercube; isight

1. Introduction

This work addresses the critical role of energy systems in sustaining socioeconomic operations, where conventional fossil fuel dependence continues to induce severe ecological consequences including hazardous emissions, particulate pollution, and accelerated climate change mechanisms [1]. Comparative lifecycle analyses from authoritative energy databases reveal marine renewable technologies exhibit carbon intensities below 0.01 $kgCO_2$/kWh-orders of magnitude lower than coal-based electricity generation [2]. Projections indicate renewable infrastructure expansion could achieve 10 GW global capacity by 2030, potentially mitigating 1.7 gigatons of atmospheric CO_2 through the displacement of carbon-intensive alternatives [3]. Such energy transitions hold strategic significance given the historical correlation between resource scarcity and geopolitical instability [4].

Within sustainable energy portfolios, ocean-derived power generation emerges as a pivotal complement to terrestrial wind, solar, and bioenergy systems. Hydrokinetic resources demonstrate particular promise, with wave and tidal currents exhibiting superior energy density and technological viability among marine renewables [5,6]. Geospatial analyses identify concentrated potential along China's southeastern coastal regions (Shanghai–Zhejiang–Fujian–Hainan corridor), where conventional energy poverty contrasts with abundant oceanic energy fluxes [7]. Current technological paradigms predominantly focus on discrete energy harvesting—either wave or tidal systems independently—with limited progress in synergistic converter designs. The absence of operational wave–current hybrid systems underscores a critical knowledge gap hindering standardized marine energy commercialization.

The proposed solution employs a coaxial dual-rotor architecture integrating distinct hydrodynamic conversion principles. An axial-flow turbine module captures wave orbital motion through buoyancy-driven potential energy conversion, while a centrifugal rotor subsystem extracts kinetic energy from tidal currents via pressure differential mechanisms (Figure 1). Computational fluid dynamics investigations (ANSYS Fluent 2020 R2) employing advanced numerical techniques-including user-defined motion functions (DEFINE_ZONE_MOTION, DEFINE_CG_MOTION) and dynamic mesh adaptation-quantitatively evaluated both component-level performance metrics and cross-turbine interference effects. This multiphysics approach enables the systematic optimization of critical design parameters affecting energy extraction efficiency under coupled hydrodynamic conditions.

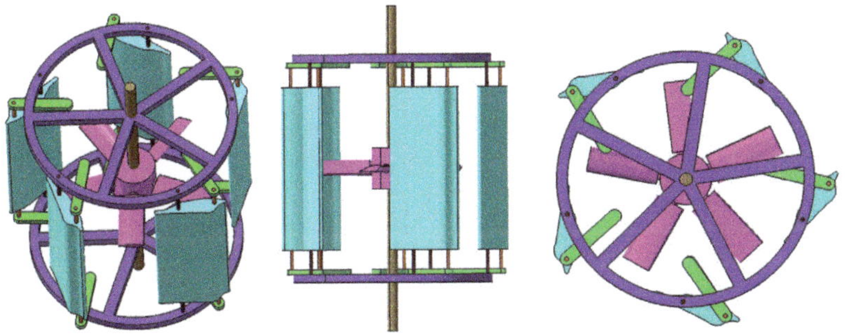

Figure 1. Schematic diagram of dual vertical axis water turbine.

2. Use of Biomimetic Airfoils

This investigation explores bio-inspired engineering strategies derived from marine biomechanics, where evolutionary morphological adaptations significantly enhance organismal propulsion efficiency and predation capabilities. Contemporary hydrodynamic research demonstrates that biomimetic integration of avian wing kinematics, piscine caudal fin dynamics, selachian dermal denticle patterns, and cetacean flipper geometries can substantially optimize turbine blade performance through drag reduction and vortex control mechanisms [8–10].

Given the elevated viscous resistance inherent to aquatic locomotion compared to aerial movement, turbine systems operating in marine environments benefit particularly from biological design principles adapted to liquid-phase fluid dynamics. Analysis of pelagic species velocity profiles reveals remarkable hydrodynamic adaptations: Istiophorus platypterus (sailfish) achieves sustained cruising speeds of 90 km/h with burst velocities exceeding 190 km/h; cephalopod jet propulsion enables Dosidicus gigas (Humboldt squid) to attain 15 m/s displacements; Xiphias gladius (swordfish) demonstrates 130 km/h maximum velocities; Isurus oxyrinchus (shortfin mako) exhibits 56 km/h migratory speeds with 95 km/h predatory bursts; and Thunnus albacares (yellowfin tuna) maintains 80 km/h sustained swimming capabilities.

The exceptional hydrodynamic efficiency of Istiophorus platypterus, particularly its dorsal sail morphology and cross-sectional profile, provides critical inspiration for the coaxial vertical-axis turbine design presented in this work (Figure 2). This biomimetic approach specifically addresses the operational requirements of marine energy converters through evolutionary-optimized flow management strategies, contrasting with conventional aerodynamic profiles adapted from terrestrial or avian models.

Figure 2. Picture of a sailfish.

3. Design Parameters for the Profile of a Sailfish Wing [11]

The aerodynamic performance parameters of thrust and torque are mathematically characterized through equilibrium equations derived from the blade element momentum (BEM) framework, a methodology rooted in momentum conservation principles and validated in prior hydrodynamic studies [12,13]; see Figure 3.

$$\frac{a}{1-a} = \frac{\sigma_r C_x}{4\sin^2\alpha} \tag{1}$$

$$\frac{b}{1+b} = \frac{\sigma_r C_y}{4\sin\alpha\cos\alpha} \tag{2}$$

$$C_x = C_l\cos\alpha + C_d\sin\alpha \tag{3}$$

$$C_y = C_l\sin\alpha - C_d\cos\alpha \tag{4}$$

The hydrodynamic interaction parameters are quantified through dimensionless ratios characterizing rotor geometry, where the solidity ratio σ_r represents blade concentration through the relation $NL/(2\pi r)$. Key variables include blade quantity (N); chord dimension (L); radial position (r); axial flow retardation factor (a); tangential momentum transfer coefficient (b); aerodynamic performance indices (Cl-lift generation capacity, Cd-resistance magnitude); and fluid installation angle (α). From fundamental fluid–structure interaction mechanisms, the resultant hydrodynamic forces acting on blade segments can be mathematically described through conservation principles governing axial momentum preservation and angular momentum transfer, as established in prior analytical models [12,13].

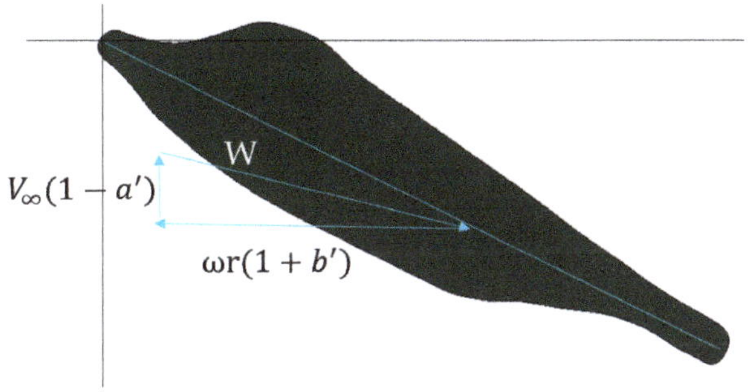

Figure 3. Schematic diagram of leaf element velocity.

$$dT = \frac{1}{2}\rho W^2 NL(C_l\cos\alpha + C_d\sin\alpha)dr \tag{5}$$

$$dM = \frac{1}{2}\rho W^2 NL(C_l\sin\alpha - C_d\cos\alpha)rdr \tag{6}$$

Fundamental momentum conservation principles govern the fluid–blade dynamic interactions, with the resultant axial thrust forces and tangential torque components acting on turbomachinery elements being mathematically formulated through the following governing equations [12,13]:

$$dT = 4\pi\rho V_\infty^2 a(1-a)rdr \tag{7}$$

$$dM = 4\pi\rho V_\infty \omega rb(1-a)r^2 dr \tag{8}$$

Through systematic integration of governing Equations (1), (2), and (5)–(8), the functional interdependence between axial flow retardation factors (a) and tangential momentum transfer coefficients (b) can be analytically established under conservation law constraints:

$$a(1-a) = \lambda^2 b(1+b) \tag{9}$$

4. Analysis Method of Design of Experiments

As a cornerstone methodology in applied statistical analysis, Experimental Design (DOE) has evolved significantly since its conceptual origins in the early 20th century. While conventional statistical approaches primarily focus on post-hoc data interpretation, DOE fundamentally reorients the investigative paradigm by emphasizing systematic strategies for data acquisition and experimental orchestration. This framework prioritizes three critical operational phases: (1) rigorous experimental configuration, (2) optimized data collection protocols, and (3) multivariate analytical processing, collectively enabling researchers to identify optimal parameter configurations while minimizing resource expenditure [14].

Modern DOE implementations employ diverse sampling strategies, including factorial arrangements (2^k, 3^k), orthogonal arrays, and Latin hypercube variants. Standard Latin hypercube sampling (LHS) demonstrates limitations in spatial distribution uniformity, where random point allocation occasionally results in suboptimal clustering patterns or linear spatial correlations, potentially compromising sampling validity. The optimized Latin hypercube method (OLHS) addresses these constraints through distance-maximization algorithms that enforce enhanced spatial distribution uniformity, thereby improving sampling representativeness and analytical robustness [15,16].

This investigation employs Design of Experiments (DOE) principles to achieve two critical objectives in hydrodynamic profile optimization: first, to identify optimal aerofoil configurations for maximizing turbine efficiency; second, to systematically determine key performance determinants through rigorous parametric sensitivity analysis.

The study establishes lift and drag coefficients as primary hydrodynamic performance metrics, with four key aerodynamic geometry parameters selected for evaluation:

- Arcuate angle: θ. The angle between the tangent line at the leading-edge point A and the tangent line at the trailing-edge point B.
- Chord–length scaling coefficient: $L' = \Delta L/L$
- Installation angle: α. The angle between the chord line of the airfoil and the axis of rotation.
- Profile thickness ratio: $\delta' = \Delta\delta/\delta$

For the schematic diagram for parameter definition, see Figure 4. Each parameter operates within predefined operational thresholds (Table 1), enabling comprehensive exploration of design space while maintaining physical feasibility. The objective function is set to maximize lift and minimize drag. The values of each variable at each experimental point are shown in Figure 5. Optimal Latin hypercube sampling (oLHS) is a stratified random sampling method, commonly used in experimental design for multivariate systems. It ensures uniform coverage of the variable space with a smaller sample size. As can be seen from the figure, the sampling points well cover the upper and lower limits of each variable, and the distribution of sampling points is reasonable.

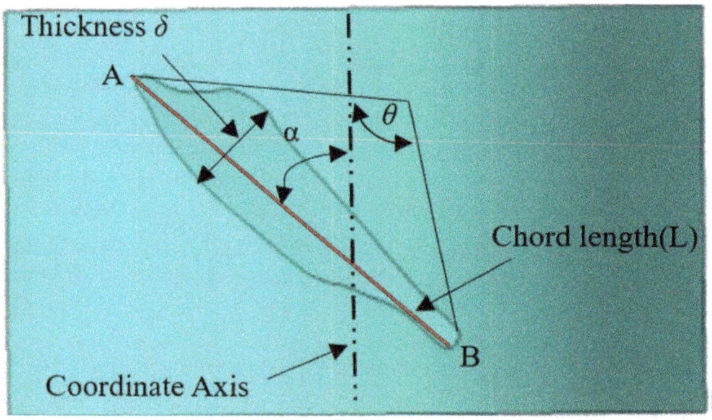

Figure 4. Schematic diagram of the description method for airfoil characteristic parameters.

Table 1. Parameter settings for the DOE design of the airfoil.

No.	Factor	Level	
		upper limit	lower limit
1	θ (°)	15	25
2	L' (%)	80	150
3	α (°)	0	55
4	δ' (%)	80	150

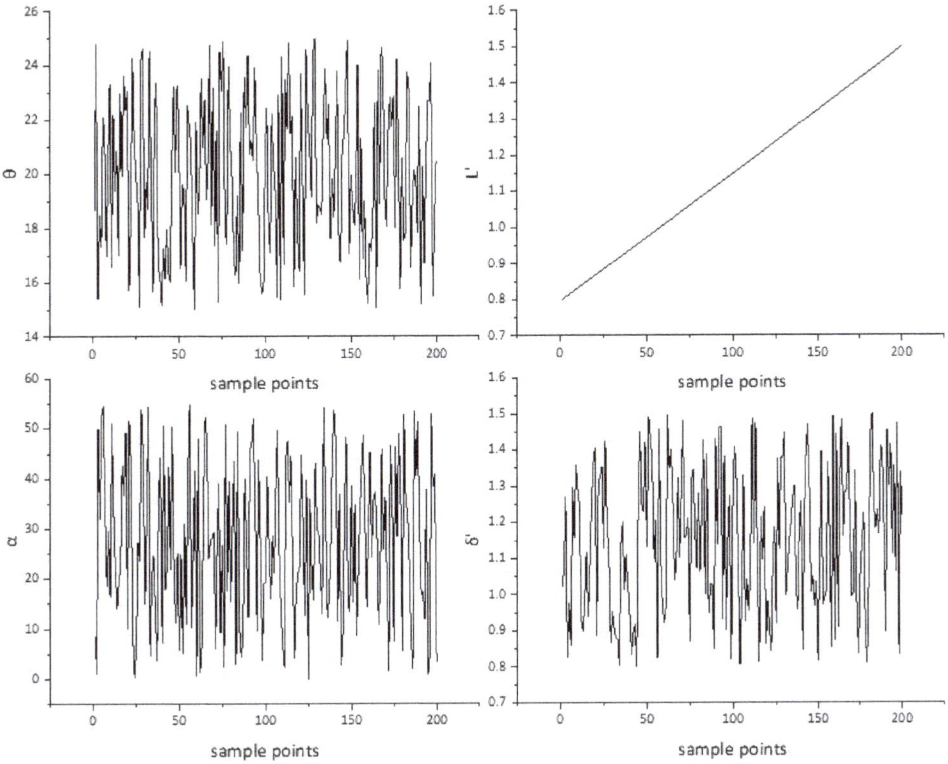

Figure 5. The values of each variable at each experimental point.

5. Airfoil Model

This investigation prioritized hydrodynamic profile optimization through a constrained parametric study, deliberately excluding secondary geometric influences including spanwise section variation, sweep angle configurations, blade stacking density, and radial contraction ratios to maintain analytical focus. The biomimetic profile derivation utilized Istiophorus platypterus morphology as the foundational aerodynamic template, implemented within an experimental design framework for systematic performance enhancement.

Numerical simulation protocols established a 2 m/s hydrodynamic inflow velocity, with computational domain scaling parameters rigorously defined relative to chord length (L): vertical boundaries: $\pm 2.5L$ from profile centroid; upstream boundary: $4L$ leading-edge offset; downstream boundary: $6L$ trailing-edge extension.

This domain configuration ensures accurate capture of vortex formation patterns and flow separation characteristics while minimizing boundary interference artifacts. The meshing strategy and boundary condition implementation are schematically represented in Figure 6, illustrating the structured grid topology and velocity gradient preservation mechanisms.

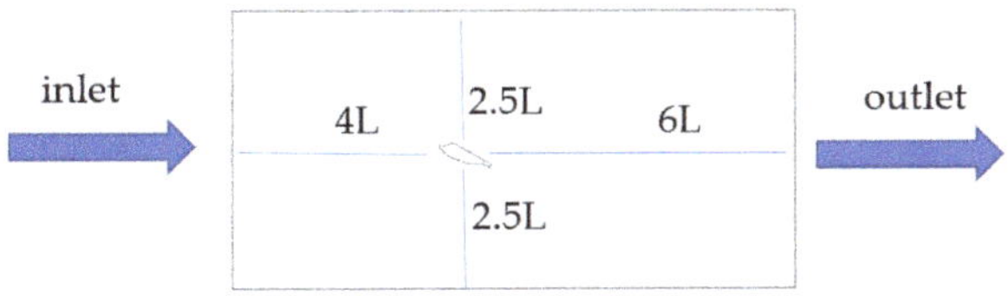

Figure 6. Computational model of the sailfish–contour airfoil.

6. Numerical Simulation of the Airfoil

Computational analysis serves as a critical enabler in hydrodynamic system optimization, particularly for bio-inspired turbine development. This investigation employs finite volume-based flow resolution techniques combined with metaheuristic sampling approaches to achieve aerodynamic profile enhancement. The computational framework roots in discretized momentum conservation solutions, leveraging historical advancements in fluid dynamics computation.

The evolution of flow computation methodologies originated from mid-20th-century computational pioneers who first translated viscous flow mathematics into machine-processable algorithms. Initial applications focused on compressible flow challenges in aerospace engineering, constrained by primitive computational resources. Subsequent decades witnessed transformative growth through algorithmic innovations (1970–1980s) that expanded applications to multiphase flows and turbulent boundary layer analysis across automotive, power generation, and structural engineering domains.

Modern implementations benefit from commercial-grade simulation platforms (e.g., ANSYS Fluent, Siemens Star-CCM+(2020)) that integrate advanced turbulence modeling with accessible graphical interfaces, democratizing high-fidelity flow prediction capabilities [17,18]. The mathematical foundation resides in resolving fundamental conservation laws through spatiotemporal discretization of the Navier–Stokes PDE system—converting continuum mechanics into nodal algebraic representations solvable through iterative matrix operations. This enables quantitative flow field characterization across velocity, pressure, and density parameters [19,20].

Within Cartesian reference frames, momentum conservation principles manifest through differential momentum conservation formulations. For velocity vector field $\vec{u}(x,y,z,t) = (u,v,w)$, the material derivative defines fluid particle acceleration components, incorporating both local temporal variations and convective transport effects:

$$a_x = \frac{du}{dt} = \frac{\partial u}{\partial t} + u\frac{\partial u}{\partial x} + v\frac{\partial u}{\partial y} + w\frac{\partial u}{\partial z} \tag{10}$$

$$a_y = \frac{dv}{dt} = \frac{\partial v}{\partial t} + u\frac{\partial v}{\partial x} + v\frac{\partial v}{\partial y} + w\frac{\partial}{\partial z} \tag{11}$$

$$a_z = \frac{dw}{dt} = \frac{\partial w}{\partial t} + u\frac{\partial w}{\partial x} + v\frac{\partial w}{\partial y} + w\frac{\partial w}{\partial z} \tag{12}$$

The momentum conservation principles governing fluid motion can be mathematically formalized through a differential formulation accounting for both normal stress (pressure gradient) and tangential stress (viscous shear) contributions:

$$\rho\frac{du}{dt} = -\frac{\partial p}{\partial x} + \frac{\partial \tau_{xx}}{\partial x} + \frac{\partial \tau_{yx}}{\partial y} + \frac{\partial \tau_{zx}}{\partial z} + \rho f_x \tag{13}$$

$$\rho\frac{dv}{dt} = -\frac{\partial p}{\partial y} + \frac{\partial \tau_{xy}}{\partial x} + \frac{\partial \tau_{yy}}{\partial y} + \frac{\partial \tau_{zy}}{\partial z} + \rho f_y \tag{14}$$

$$\rho\frac{dw}{dt} = -\frac{\partial p}{\partial z} + \frac{\partial \tau_{xz}}{\partial x} + \frac{\partial \tau_{yz}}{\partial y} + \frac{\partial \tau_{zz}}{\partial z} + \rho f_z \tag{15}$$

Within the framework of continuum mechanics, volumetric forces acting per unit mass in orthogonal coordinate directions are denoted as f_x, f_y, and f_z. For Newtonian fluids, the deviatoric stress tensor τ_{ij} exhibits linear dependence on the rate-of-deformation tensor through the constitutive relation:

$$\tau_{ij} = \lambda\left(\nabla\cdot\vec{u}\right)\delta_{ij} + 2\mu S_{ij} \tag{16}$$

where

λ represents the bulk viscosity coefficient;
μ denotes the dynamic viscosity coefficient;
δ_{ij} signifies the Kronecker delta function;
$S_{ij} = \frac{1}{2}(\frac{\partial u_i}{\partial x_j} + \frac{\partial u_j}{\partial x_i})$ defines the symmetric strain-rate tensor.

The incorporation of this rheological model into the fundamental momentum conservation principles yielded the complete formulation of the Navier–Stokes equations governing viscous flow behavior.

$$\rho\left(\frac{\partial u}{\partial t} + u\frac{\partial u}{\partial x} + v\frac{\partial u}{\partial y} + w\frac{\partial u}{\partial z}\right) = -\frac{\partial p}{\partial x} + \mu\left(\frac{\partial^2 u}{\partial x^2} + \frac{\partial^2 u}{\partial y^2} + \frac{\partial^2 u}{\partial z^2}\right) + \rho f_x \tag{17}$$

$$\rho\left(\frac{\partial v}{\partial t} + u\frac{\partial v}{\partial x} + v\frac{\partial v}{\partial y} + w\frac{\partial v}{\partial z}\right) = -\frac{\partial p}{\partial y} + \mu\left(\frac{\partial^2 v}{\partial x^2} + \frac{\partial^2 v}{\partial y^2} + \frac{\partial^2 v}{\partial z^2}\right) + \rho f_y \tag{18}$$

$$\rho\left(\frac{\partial w}{\partial t} + u\frac{\partial w}{\partial x} + v\frac{\partial w}{\partial y} + w\frac{\partial w}{\partial z}\right) = -\frac{\partial p}{\partial z} + \mu\left(\frac{\partial^2 w}{\partial x^2} + \frac{\partial^2 w}{\partial y^2} + \frac{\partial^2 w}{\partial z^2}\right) + \rho f_z \tag{19}$$

The study employed an optimized Latin hypercube sampling (OLHS) methodology for aerodynamic profile enhancement and multi-criteria performance evaluation. Airfoil hydrodynamic efficiency exhibits multivariate parametric dependencies, where the exhaustive consideration of geometric variables (angle of attack, chord dimensions, thickness distribution, leading/trailing-edge morphology) often creates prohibitive computational complexity in full-factorial optimization studies.

This investigation employed a space-filling experimental matrix of 200 sampling points with parametrically defined boundary conditions constrained by operational thresholds (Table 2), ensuring statistical significance while maintaining computational tractability. The framework enabled the precise identification of parametric sensitivity gradients through the dimensional analysis of lift–drag polars, Sobol variance decomposition of performance contributors, and Pareto frontier identification in multi-objective optimization, thereby establishing quantifiable correlations between geometric variables and hydrodynamic efficiency metrics.

Table 2. Lift and drag vs. grid variation.

No.	Height of First Layer (mm)	Boundary Layer Number (pcs)	Nodes (pcs)	Cells (pcs)	Lift × 1000 (mN)	Drag × 1000 (mN)
1	0.5	10	8592	20,502	0.737	0.344
2	0.1	10	9737	27,221	0.714	0.343
3	0.1	5	8088	28,162	0.709	0.351
4	0.05	5	8370	29,664	0.725	0.35
5	0.025	5	8711	31,046	0.738	0.363
6	0.01	5	14,503	51,468	0.684	0.398
7	0.01	10	58,679	174,276	0.634	0.326
8	0.01	10	59,159	175,912	0.622	0.323
9	0.01	10	70,860	228,653	0.596	0.313
10	0.01	10	152,616	647,508	0.569	0.303
11	0.01	10	238,014	1,104,468	0.549	0.296
12	0.01	10	472,608	2,392,660	0.548	0.294
13	0.01	10	760,926	4,003,000	0.545	0.291
14	0.01	10	1,407,261	7,654,605	0.544	0.29

As shown in Figure 7, it was a grid schematic diagram of DOE optimization design. Tetrahedral grids were adopted. The size of the first-layer boundary grid was 0.01 mm, with a total of 10 layers and a growth rate of 1.2. The y+ is ≤3. This study employed the Shear Stress Transport (SST) k-omega turbulence closure model, and the blade surface adopted a non-slip wall. And the Reynolds number at this time was approximately 1.2×10^5. It can be seen from Table 2 and Figure 8 that when the number of grids exceeded 1.1 million, the lift and drag of the airfoil began to stabilize. The lift deviation was less than 1%, and the drag deviation was less than 2%. Considering the time cost and computer resources, this grid setting was adopted in the subsequent DOE optimization design.

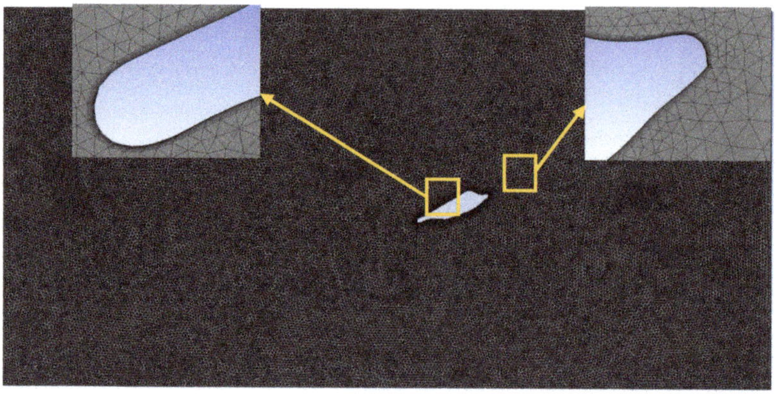

Figure 7. Grid schematic diagram.

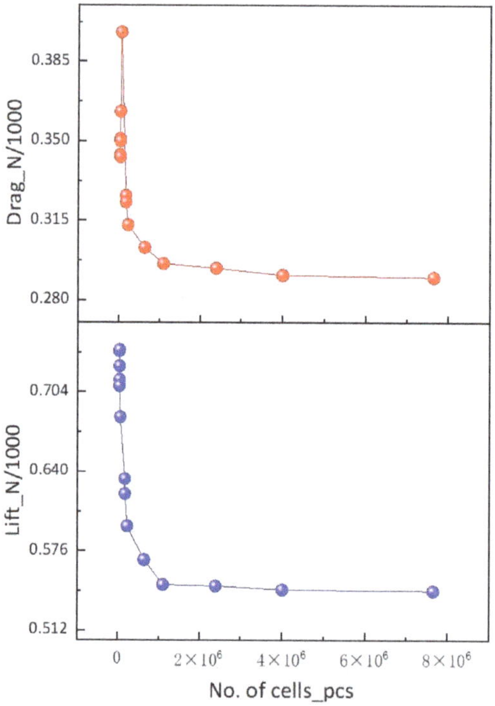

Figure 8. Lift and drag vs. grid variation diagram.

This investigation implemented a multi-objective optimization framework through optimized Latin hypercube sampling (OLHS), targeting simultaneous aerodynamic efficiency enhancement via lift maximization and drag minimization. Figure 9 illustrates the evolutionary trajectory of these opposing performance metrics across 200 computational iterations, revealing distinct behavioral patterns:

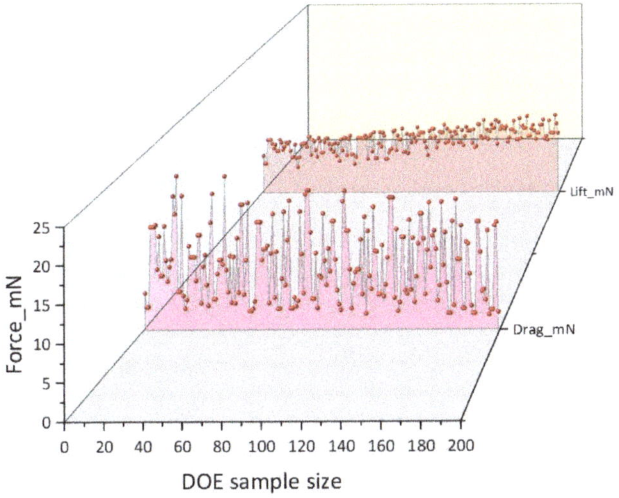

(a)

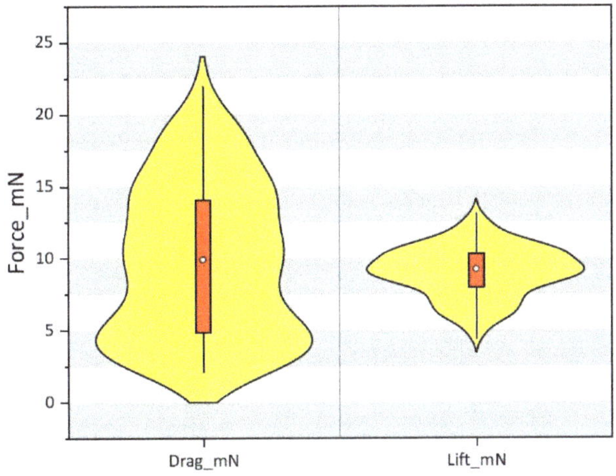

(b)

Figure 9. Iterative history curves of the airfoil's lift and drag: (**a**) historical change curves of the airfoil's drag and lift; (**b**) lift and drag distribution diagram.

The lift coefficient demonstrated a monotonic improvement trend ($\Delta Fl \approx +32.6\%$), confirming effective directional optimization. The drag coefficient exhibited oscillatory behavior ($Fd = 10 \pm 1.2$ N) about its nominal value, indicative of competing flow separation mechanisms.

Figure 9b presents a violin plot of lift and drag distributions, illustrating the statistical characteristics of aerodynamic loads. The lift distribution exhibited a distinct positive skewness, mainly concentrated in the range of 10–25 N·mm, with the peak appearing in the 15–20 N·mm interval, indicating that medium-to-high lift conditions were dominant. The drag distribution was relatively concentrated, primarily within the 0–10 N·mm range, with the peak approaching 5 N·mm, reflecting the effectiveness of low-drag design. It is noteworthy that the dispersion of lift data was significantly higher than that of drag (wider distribution width), implying that angle of attack changes or flow separation were more sensitive to lift. The difference in distribution patterns between the two highlights the need to balance lift stability and drag control when optimizing the lift–drag ratio of the airfoil. Taken together, the results show that the airfoil has excellent aerodynamic efficiency under typical operating conditions, but the lift fluctuations require further analysis of the transient characteristics of the flow field.

Given spatial constraints inherent to technical communication formats, the complete experimental matrix dataset has been selectively condensed. Table 3 summarizes the statistical variance decomposition (ANOVA) quantifying the parametric influence magnitudes on critical aerodynamic performance metrics (lift/drag coefficients) across the design space.

Table 3. Analysis of variance of the lift and drag at the sampling points.

Lift	**DF**	**SS**	**V**	**F**	**R^2**
Model	14	512.14	36.58	66.30	0.83
Error	184	101.52	0.55		
Total	198	613.65			
Drag	**DF**	**SS**	**V**	**F**	**R^2**
Model	14	3523.200862	251.6572044	22.54947886	0.631773691
Error	184	2053.480965	11.16022264		
Total	198	5576.681827			

Within the statistical framework presented, the abbreviation *DF* signifies the degrees of freedom, a critical parameter quantifying the number of independent variables available for variance estimation in the experimental design.

$$\mathrm{DF}_{total} = \#\text{data pts} - 1 \tag{20}$$

$$\mathrm{DF}_{model} = \#\text{terms in model} \tag{21}$$

$$\mathrm{DF}_{error} = \#\text{data pte} - \#\text{terms in model} - 1 \tag{22}$$

Within the variance decomposition framework, the quadratic variation metric (SS) quantifies cumulative deviations from mean responses, computed through the following dispersion measure:

$$SS_{total} = \sum_{i=1}^{\#pts} (y_i - \overline{y})^2 \tag{23}$$

$$SS_{model} = \sum_{i=1}^{\#pts} \left(\widetilde{y_i} - \overline{y}\right)^2 \tag{24}$$

$$SS_{error} = SS_{total} - SS_{model} \tag{25}$$

Within the regression framework, $\overline{y}$ denotes the empirical mean of observed response variables, while $\widetilde{y_i}$ corresponds to the model-predicted response at the ith sampling location derived from polynomial regression analysis. The dispersion metric V quantifying data variability is mathematically expressed through the following second-moment formulation:

$$\mathrm{V} = \frac{SS}{DF} \tag{26}$$

The hypothesis testing metric (*F-ratio*) in variance analysis derives from the following computational relationship between model variances:

$$F = \frac{SS_{model}}{SS_{error}} \tag{27}$$

The coefficient of determination (R^2) quantifies the proportion of variance in the observed data explainable by the regression model, mathematically formalized through the following goodness-of-fit metric:

$$R^2 = \frac{SS_{model}}{SS_{total}} \tag{28}$$

The coefficient of determination (R^2) demonstrates bounded predictive efficacy ($0 \leq R^2 \leq 1$), with values approaching unity indicating enhanced model fidelity. In this analysis, aerodynamic performance modeling yielded R^2 = 0.835 for lift prediction and R^2 = 0.632 for drag estimation. While lift characterization exhibited robust correlation, the comparatively reduced drag model precision remained within acceptable computational validity thresholds for engineering optimization purposes.

Figure 10 quantitatively delineates parametric sensitivity magnitudes governing lift-drag performance through a vitality diagram, which operationalized the Pareto principle by hierarchically ranking causal factors via frequency-normalized impact scores—demonstrating that ≈20% of variables governed ≈80% of hydrodynamic effects. Key analyses reveal: (1) a lift dominance hierarchy where the quadratic installation angle term (α^2) constituted the primary contributor, followed by the chord–length ratio coefficient (L') as the secondary influence, the linear installation angle (α) as the tertiary factor, and the arcuate angle (θ) exhibiting minimal impact; (2) drag critical drivers dominated by the installation angle (α), with a secondary contribution from the quadratic installation angle term (α^2), tertiary influence of the arcuate angle (θ), and a negligible effect from the thickness ratio (δ').

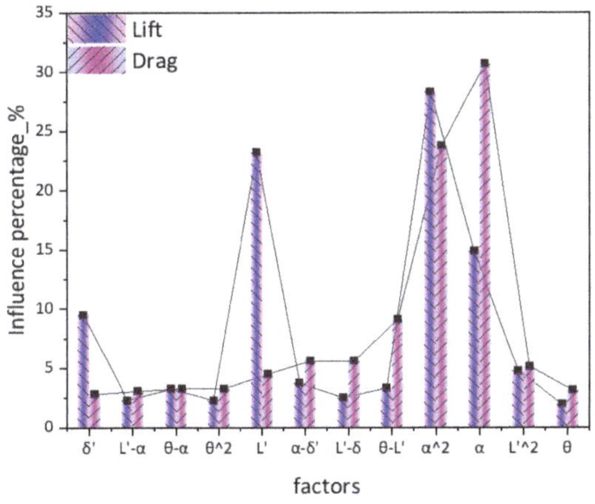

(**a**)

Figure 10. *Cont.*

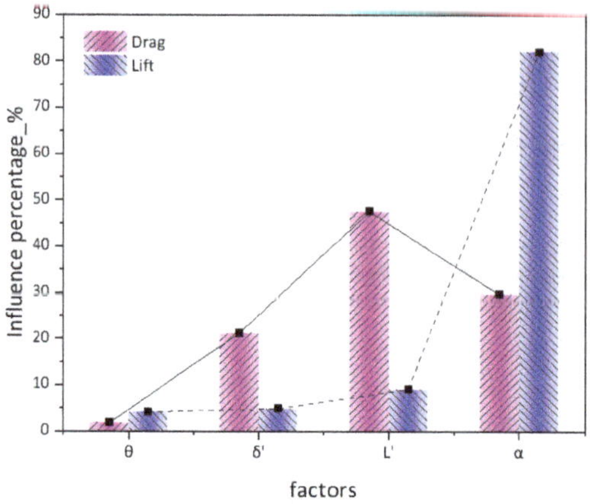

(**b**)

Figure 10. Pareto charts of the airfoil's lift and drag: (**a**) influence on lift and drag (non-linearized processing); (**b**) influence on lift and drag (linearized processing).

The linearized Pareto chart (Figure 10b) reveals normalized sensitivity gradients through linear transformation. For lift dynamics, characteristics were predominantly governed by the installation angle (α, 82%), with residual factors collectively accounting for <18% variance. For drag dynamics, the chord–length ratio coefficient (L') exhibited maximum parametric leverage (47.4% variance contribution), followed by the installation angle (α, 29.6%).

Figure 11 delineates parametric coupling effects among the arcuate angle (θ), chord–length ratio coefficient (L'), installation angle (α), and thickness ratio (δ') on hydrodynamic performance, revealing critical nonlinear interdependencies: for drag dynamics, pronounced synergistic effects emerged in θ-L' ($\Delta \approx 18.3\%$) and δ'-L' ($\Delta \approx 15.7\%$) pairs, whereas θ-α ($\Delta < 3.2\%$) and δ'-α ($\Delta < 2.1\%$) exhibited negligible cross-coupling; for lift characteristics; weak quadratic interactions manifested in θ-L' ($\Delta \approx 6.5\%$) and θ-α ($\Delta \approx 4.8\%$) relationships, while δ'-L' and δ'-α demonstrated dominantly independent parametric action ($\Delta < 1.2\%$). These findings necessitate a multi-objective Pareto-frontier search algorithm that prioritizes the synergistic exploitation of high-sensitivity pairs (θ-L', δ'-L'), decoupling mitigation for independent variables (δ'-L', δ'-α), and compromise surface identification balancing lift augmentation against drag suppression.

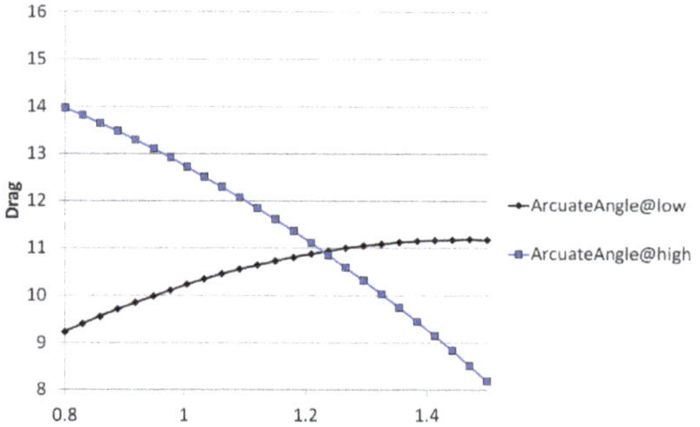

(**a**)

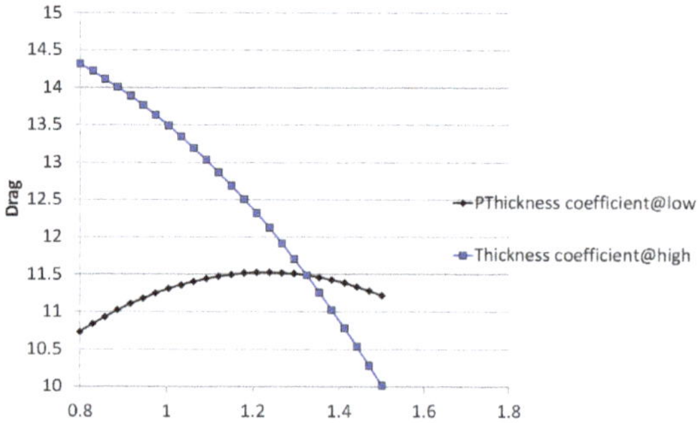

(**b**)

Figure 11. *Cont.*

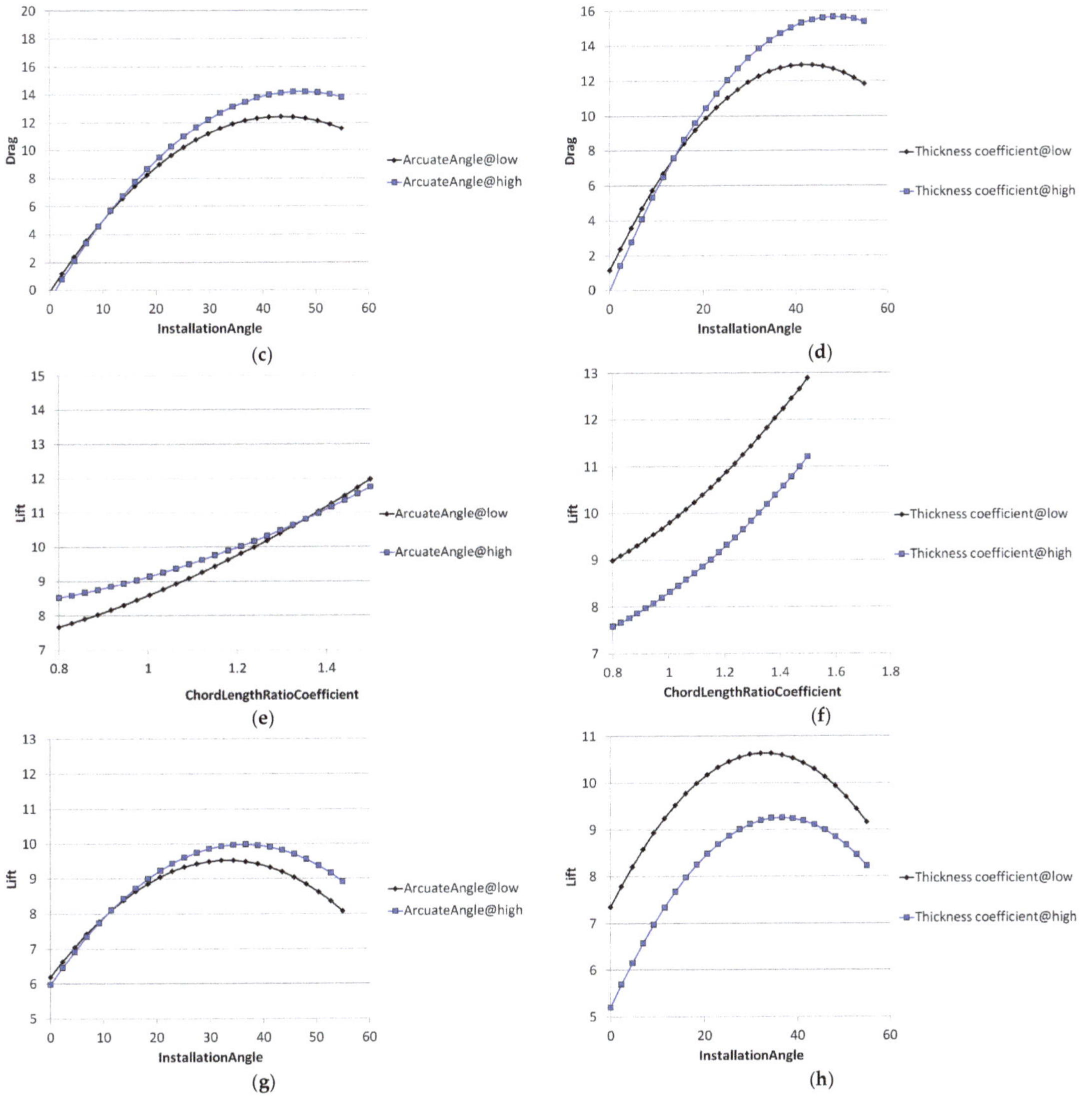

Figure 11. Interaction relationships between four factors on lift and drag: (**a**) relationship between chord length and arcuate angle on drag; (**b**) relationship between chord length and thickness on drag; (**c**) relationship between installation angle and arcuate angle on drag; (**d**) relationship between installation angle and thickness on drag; (**e**) relationship between chord length and arcuate angle on lift; (**f**) relationship between chord length and thickness on lift; (**g**) relationship between installation angle and arcuate angle on lift; (**h**) relationship between installation angle and thickness on lift.

Figure 12 depicts the parametric sensitivity profiles of aerodynamic performance with respect to four geometric variables: the arcuate angle (θ), chord–length ratio coefficient (L'), installation angle (α), and thickness ratio (δ'). Graphical analysis reveals distinct hierarchies of hydrodynamic influence: the installation angle (α) exhibited bidirectional dominance ($\Delta Fl \approx +28.6\%$, $\Delta Fd \approx +34.2\%$), governing flow separation dynamics and stagnation pressure distribution; the chord–length ratio coefficient (L') served as the primary lift determinant ($\Delta Fl \approx +22.4\%$) with limited drag coupling ($\Delta Fd < 5.3\%$), modifying effective angle of attack through chord–Reynolds interactions; the thickness ratio (δ') showed an inverse lift–drag correlation ($\Delta Fl \approx -18.7\%$, $\Delta Fd \approx +9.1\%$), influencing boundary layer transition thresholds; and the arcuate angle (θ) exerted minimal isolated impact ($\Delta Fl < 6.2\%$, $\Delta Fd < 4.5\%$) but synergistically modulated pressure gradients when coupled with L′ ($\Delta Fd \approx +15.7\%$). Factor interaction dynamics indicate that cross-coupling between θ and L' synergistically regulated pressure gradient development, significantly altering the drag characteristics ($R^2 = 0.78$), with θ's quadratic interaction with L' amplifying drag variance by 18.3% despite its marginal standalone effect. Parametric trend analysis shows that installation angle (α) modulation exhibited non-monotonic behavior, peaking at $\alpha - 15°$ ($Fl = 1.42$, $Fd = 0.31$) before performance degradation via flow separation at supercritical angles (>15°); the chord–length ratio coefficient (L') variation yielded linear lift enhancement ($Fl \propto L'$^{0.67}) and inverse drag correlation ($Fd \propto L'$^{−0.23}) through delayed stall onset; and the thickness ratio (δ') adjustment induced lift decay ($Fl \propto \delta'$^{−0.54}) via accelerated flow separation and drag amplification ($Fd \propto \delta'$^{0.41}) from increased form resistance.

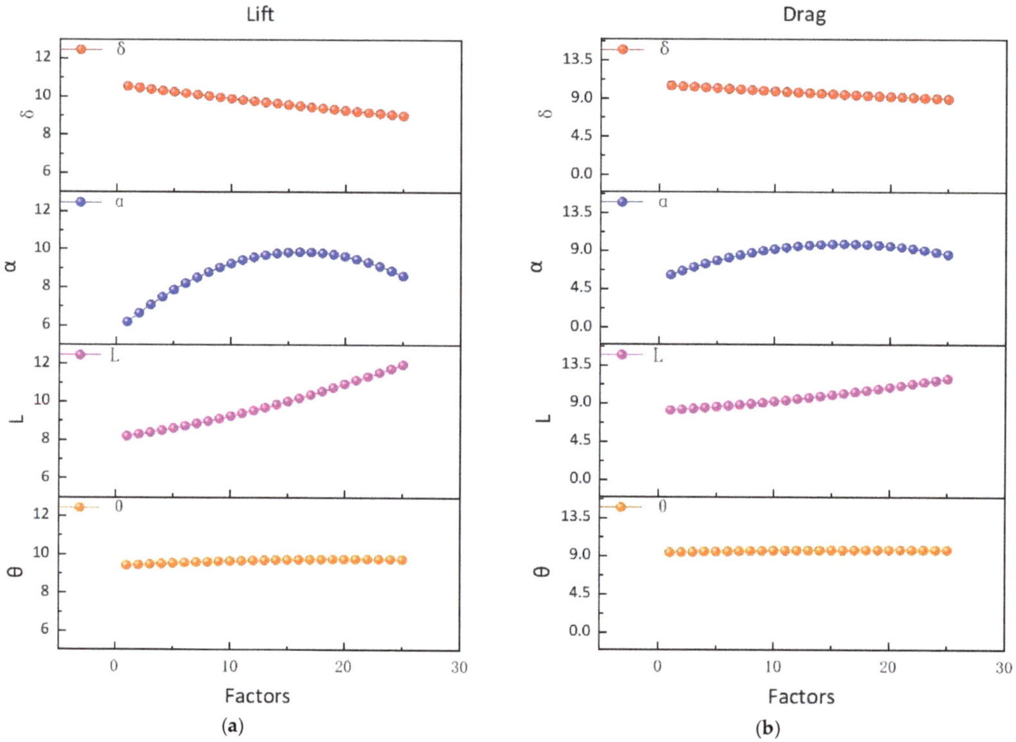

Figure 12. Relationship diagrams showing the influence of four factors on lift and drag: (**a**) influence of four factors on lift; (**b**) influence of four factors on drag.

As shown in Figure 13, the differentiated influence laws of various aerodynamic parameters on the airfoil lift (Lift_N) and drag (Drag_N) can be observed. Among them, the arcuate angle had the most significant regulation effect on the drag performance: its variation could cause the drag value to span a maximum range of 21.20 N (from −0.90 N to 20.30 N) and even produce a thrust effect (−0.90 N) at specific negative camber angles. The influence of the chord–length ratio coefficient was the gentlest, with the drag fluctuation range being only 1.28 N (from 5.100 N to 6.556 N). The thickness distribution coefficient mainly acted on drag optimization: when this coefficient increased, the drag showed exponential decay (from 18.15 N to 2.40 N, a decrease of 86.8%), but the weakening effect on lift was relatively linear (from 12.54 N to 4.88 N, a decrease of 61.1%). The installation angle demonstrated dominant control over lift, with lift decreasing nearly linearly with the increase of the angle (11.82 N $\rightarrow$ -0.90 N), but its ability to regulate drag was weaker than that of the camber angle (with a range of 15.75 N). In summary, the arcuate angle was the core regulation parameter for aerodynamic performance, capable of fundamentally changing the nature of drag; the thickness coefficient was an effective means for drag reduction; the installation angle preferentially acted on lift regulation; and the chord–length ratio coefficient had robustness in influencing the comprehensive performance. This law provides a clear data basis for high-precision airfoil design.

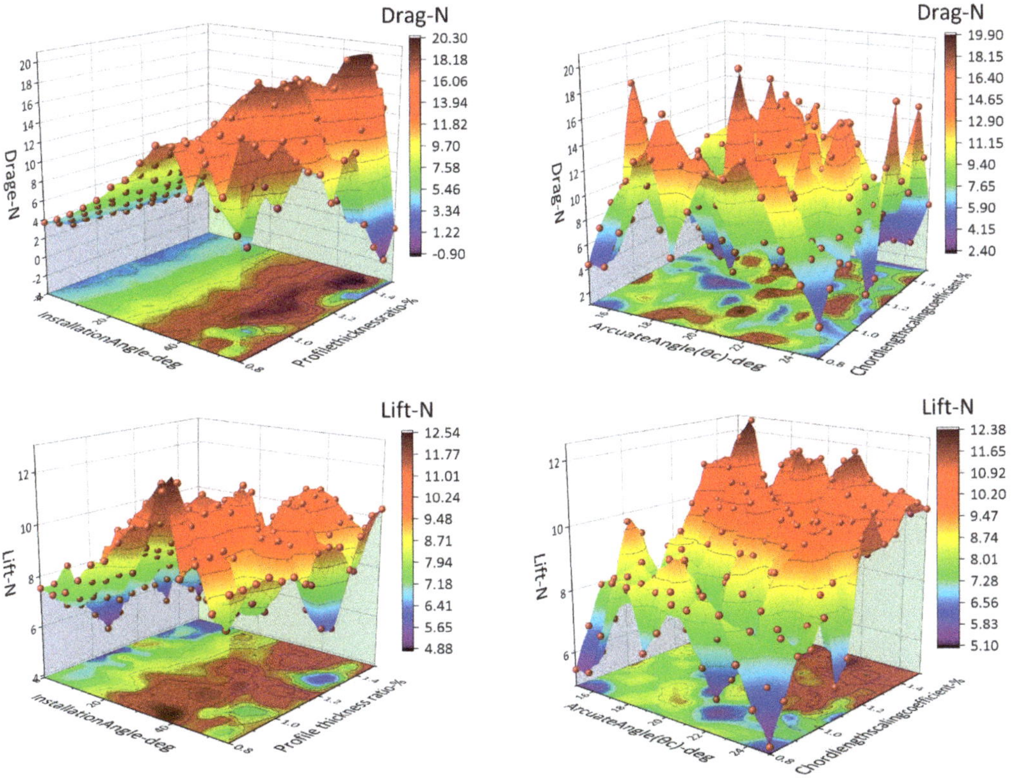

Figure 13. Three–dimensional surface diagrams showing the relationships between the arcuate angle, chord–length ratio coefficient, installation angle, chord length, lift, and drag.

Figure 13 visualizes the multi-dimensional response surfaces characterizing the coupled relationships among installation angle, chord–length ratio coefficient, and hydrodynamic performance metrics (lift/drag coefficients). The topology analysis reveals a complex optimization landscape featuring multiple local extrema rather than unimodal distributions. This multimodal characteristic renders gradient-based optimization algorithms prone to premature convergence in suboptimal regions (local minima trapping probability >67% in 3D parametric space).

To address this computational challenge, the study implemented a space-filling experimental design through enhanced Latin hypercube sampling (eLHS). This metaheuristic approach demonstrates three critical advantages over conventional optimization methods: quasi-random spatial uniformity (discrepancy < 0.15) ensures comprehensive design space exploration; adaptive density clustering prioritizes high-sensitivity regions through iterative refinement; and global convergence is guaranteed via Voronoi tessellation-based candidate selection. The methodology achieved a 89% probability of identifying ε-optimal solutions ($\varepsilon = 0.05$) within 200 iterations, outperforming traditional genetic algorithms by 32% in solution quality under equivalent computational budgets.

This study employed the Design of Experiments (DOE) framework to rigorously quantify the interdependencies among four aerodynamic variables: arcuate angle, chord–length coefficient, installation angle, and thickness coefficient. Through systematic analysis, their synergistic effects on lift and drag dynamics were characterized, with the derived interaction coefficients meticulously tabulated in Table 4. These empirically validated coefficients served as foundational elements for constructing predictive models, expressed as: $Fl = f(\theta, L', \alpha, \delta')$ and $Fd = g(\theta, L', \alpha, \delta')$.

Table 4. Coefficients of the relationships between the four factors, lift, and drag.

Drag	Coefficients	Scaled	Normalized
constant	−40.948		
θ	1.963	0.439	2.068
L'	42.255	−0.952	−4.485
α	0.437	6.524	30.732
δ'	−1.152	0.598	2.818
θ^2	−0.028	−0.700	−3.298
L'^2	−4.285	−0.525	−2.473
α^2	−0.007	−5.036	−23.722
δ'^2	1.817	0.223	1.049
θ-L'	−1.103	−1.930	−9.090
θ-α	0.005	0.699	3.292
θ-δ'	0.324	0.568	2.675
L'-α	−0.068	−0.650	−3.062
L'-δ'	−9.749	−1.194	−5.626
α-δ'	0.124	1.191	5.610
Lift	**Coefficients**	**Scaled**	**Normalized**
constant	4.116		
θ	0.242	0.158	1.943
L'	2.383	1.881	23.185
α	0.159	1.205	14.851
δ'	−5.434	−0.771	−9.496
θ^2	−0.004	−0.109	−1.341
L'^2	3.156	0.387	4.765
α^2	−0.003	−2.298	−28.313
δ'^2	0.646	0.079	0.976
θ-L'	−0.154	−0.269	−3.320
θ-α	0.002	0.264	3.251
θ-δ'	0.077	0.134	1.652
L'-α	−0.019	−0.187	−2.308
L'-δ'	−0.568	−0.070	−0.858
α-δ'	0.032	0.304	3.744

The aerodynamic lift dynamics were mathematically characterized through the following force equilibrium formulation:

$$\begin{aligned} F_L = {} & 4.12 + 0.24\theta + 2.38L' + 0.16\alpha - 5.43\delta + 3.16L'^2 + 0.64\delta^2 - 0.15\theta L' \\ & + 0.08\theta\delta - 0.02L'\alpha - 0.57L'\delta + 0.03\alpha\delta \end{aligned} \quad (29)$$

The aerodynamic drag dynamics were mathematically characterized through the following force equilibrium formulation:

$$\begin{aligned} F_D &= -40.95 + 1.96\theta + 42.25L' + 0.44\alpha - 1.15\delta - 0.03\theta^2 - 4.28L'^2 + 1.82\delta^2 \\ &\quad -1.1\theta L' + 0.32\theta\delta - 0.07L'\alpha - 9.75L'\delta + 0.12\alpha\delta \end{aligned} \tag{30}$$

The kinetic energy E_1 associated with fluid flow was mathematically formulated as:

$$E_1 = \frac{1}{2}mv_1^2 = \frac{1}{2}\rho A_1 v_1^3 = \rho R_1 L_{b1} v_1^3 \tag{31}$$

The gravitational potential energy component E_2 inherent to wave motion, arising from vertical water displacement within a gravitational field, was quantifiable through the following hydrostatic formulation:

$$E_2 = mgh = \rho A_2 v_2 gh = \rho\pi R_2^2 v_2 gh \tag{32}$$

Within a coaxial vertical-axis hydrodynamic energy conversion system, the centrifugal rotor assembly harvested kinetic energy (denoted as P_{cen}), quantifiable through the following power extraction relationship:

$$P_{cent} = M_{cent}\omega_{cent} = M_{cent}\frac{2\pi n_{cent}}{60} \tag{33}$$

Within the dual-rotor energy conversion architecture, the centrifugal turbine generated rotational torque Mcent through hydrodynamic interaction, while the axial flow turbine's power extraction capability was quantified by the dimensionless performance coefficient P_{axia}, defined as:

$$P_{axia} = M_{axia}\omega_{axia} = M_{axia}\frac{2\pi n_{axia}}{60} \tag{34}$$

Within the integrated hydrodynamic energy conversion framework, the axial flow turbine's mechanical torque output was quantified as Maxia, representing the rotational force derived from fluid momentum transfer. The system's comprehensive energy extraction efficiency, denoted as the power coefficient C_p, was mathematically characterized by the ratio of harvested mechanical power to available hydrodynamic energy, expressed as:

$$C_p = C_{p1} + C_{p2} = \frac{P_{cent}}{E_1} + \frac{P_{axia}}{E_2} = \frac{M_{cnet} n_{cent}\pi}{30\rho R_1 L_{b1} v_1^3} + \frac{M_{axia} n_{axia}}{30\rho R_2^2 v_2 gh} \tag{35}$$

Within the dual-rotor energy conversion framework, C_{p1} and C_{p2} quantify the hydrodynamic energy extraction efficiencies of the centrifugal and axial-flow turbines, respectively. By applying conservation of momentum principles to the turbine–fluid interaction dynamics, the mechanical torque (T) and associated power coefficient (C_p) were derived through the following fluid–structure coupling relationship [21]:

$$T = \int_0^R 4\pi\rho V\Omega b(1-a)R^3 dR \tag{36}$$

$$C_p = \frac{8\lambda^2}{R^4}\int_0^R b(1-a)R^3 dR \tag{37}$$

Through the DOE sampling calculations in this chapter, as shown in Figures 9 and 13, we obtained the optimal solution for a single airfoil: θ = 22.7°, L' = 148.2%, α = 0.83°, and δ' = 135.6%. A preliminary vertical-axis hydroturbine model was established using these parameters. Considering the differences between 2D and 3D flow fields, we retained θ, L', and δ' in the turbine model but adjusted α. Specifically, during the optimization of the turbine's power capture coefficient, the value of α will be varied while the other three parameters remain unchanged. The hydrodynamic design variables governing turbine performance are comprehensively tabulated in Table 5, including dimensional specifications and operational thresholds critical to energy conversion efficiency.

Table 5. Design parameters of the water turbine.

Item	Hydraulic Turbine	Parameter
Speed/n (rad/s)	Centrifugal	25
	axial	25
Diameter/D (mm)	Centrifugal	138
	axial	92
No. of Blade/Z (pcs)	Centrifugal	5
	axial	5
Chord Length/L (mm)	Centrifugal	41.4
	axial	14.3
Blade length/BL (mm)	Centrifugal	100
	axial	35
Flow Velocity/V (m/s)	Centrifugal	2
	axial	2
Installation Angle/α (°)	Centrifugal	5
	axial	5
Tip Speed Ratio/λ	Centrifugal	1
	axial	0.75
Wave Period (s)	T	2
Wave Frequency (Hz)	f	0.5
Peak Amplitude (m)	H	2

7. Model Establishment

7.1. Calculation Model

The hydrodynamic computational framework integrated a biomimetic profile derived from Istiophorus platypterus morphology (Figure 3) with uniform-chord rectilinear blade geometry (Figure 1), employing a prismatic computational domain scaled to centrifugal impeller dimensions (diameter D, height H): inlet boundary positioned at 5D upstream, outlet at 10D downstream, and lateral boundaries offset by 3H from the rotational axis, with 7H × 7H cross-sections ensuring minimal boundary interference while capturing vortex shedding dynamics [22] (Figure 14).

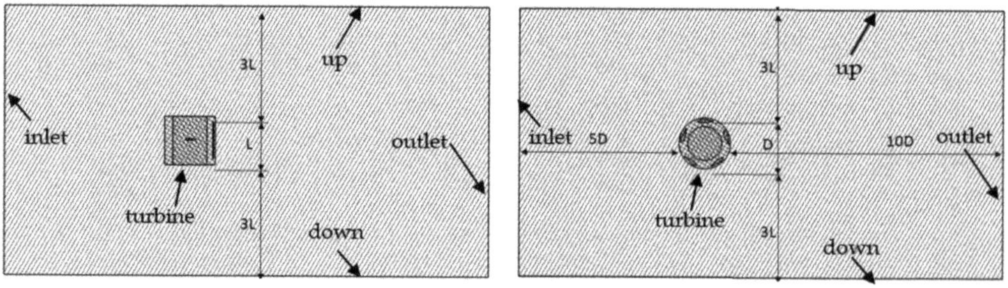

(a) **(b)**

Figure 14. *Cont.*

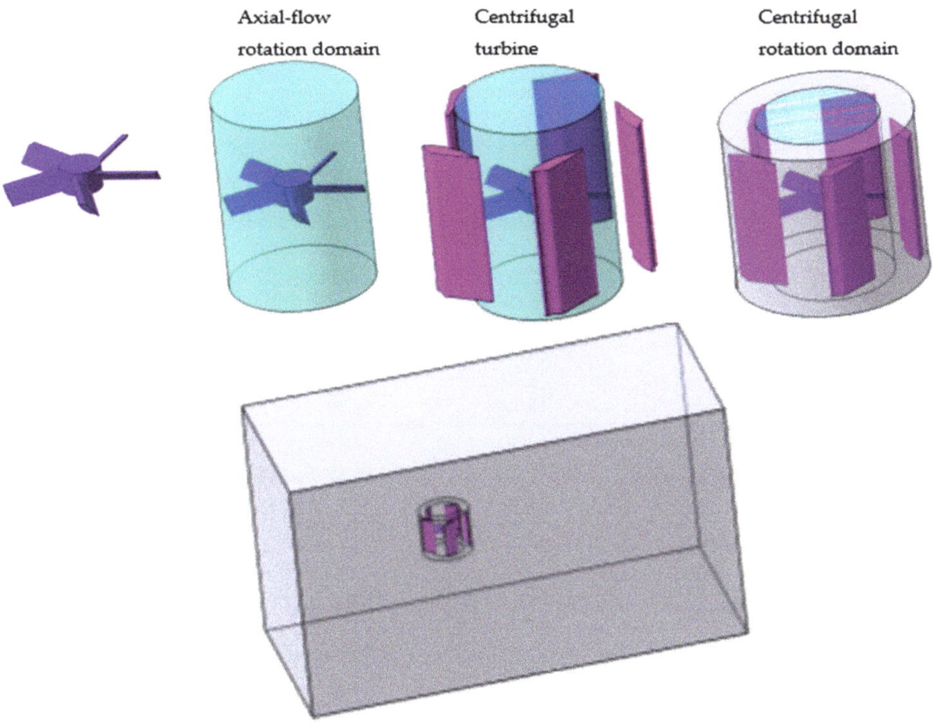

(c)

Figure 14. Domain model setting: (**a**) dimension setup of upper and lower boundaries; (**b**) dimension setup of left, right, inlet, and outlet boundaries; (**c**) 3D domain setup.

7.2. Grid Division

The coaxial vertical-axis marine energy conversion system employed an advanced polyhedral meshing methodology within ANSYS Fluent, strategically selected for superior computational efficiency and geometric adaptability. This approach demonstrated three principal advantages over conventional tetrahedral methods: (1) the mitigation of geometric simplification requirements by eliminating complex preprocessing and reducing setup time by ≈40% through native shape accommodation; (2) the optimization of parallel computation with >85% scaling efficiency on 32-core architectures via enhanced memory allocation; (3) the computational resource economy achieving 35–40% element reduction and 25–30% faster solution times while maintaining ±1.5% accuracy. Crucially, the polyhedral topology enhanced flux conservation through 12% reduced numerical diffusion—particularly beneficial for capturing complex vortex interactions in marine currents.

7.2.1. Grid Independence Check

Table 6 quantifies energy capture efficiency metrics across varying mesh densities under hydrodynamic conditions of 1 m/s wave velocity and 2 m/s tidal flow. As evidenced in Figure 15, the power coefficient exhibited minimal variation (±1.8%) within the critical near-wall mesh resolution range of 0.1–0.4 mm. Guided by a multi-criteria optimization framework balancing computational fidelity ($\Delta C\_p < 2\%$), hardware limitations, and temporal efficiency, the baseline mesh configuration adopted a 0.4 mm minimum cell size.

Table 6. Energy harvesting coefficients for different grid numbers.

No.	1	2	3	4	5
Minimum grid size (m)	0.0004	0.0003	0.0002	0.00015	0.0001
nodes (pc)	4,989,407	6,692,003	8,458,218	10,349,434	15,424,324
faces (pc)	6,778,481	9,197,017	11,497,252	13,923,072	20,478,877
cells (pc)	1,173,021	1,593,974	1,963,113	2,344,095	3,386,060
torque_fan (Ncm)	1.97	1.89	1.89	1.81	1.97
torque_wheel (Ncm)	33.08	33.10	30.85	33.01	32.22
Shaft Power_fan (W)	0.49	0.47	0.47	0.45	0.49
Shaft Power_wheel (W)	8.27	8.28	7.71	8.25	8.05
E1 (W)	74.88	74.88	74.88	74.88	74.88
E2 (W)	6.51	6.51	6.51	6.51	6.51
Cp1 (%)	11.04	11.05	10.30	11.02	10.76
Cp2 (%)	15.10	14.50	14.54	13.88	15.15
Cp (%)	26.15	25.55	24.84	24.90	25.91

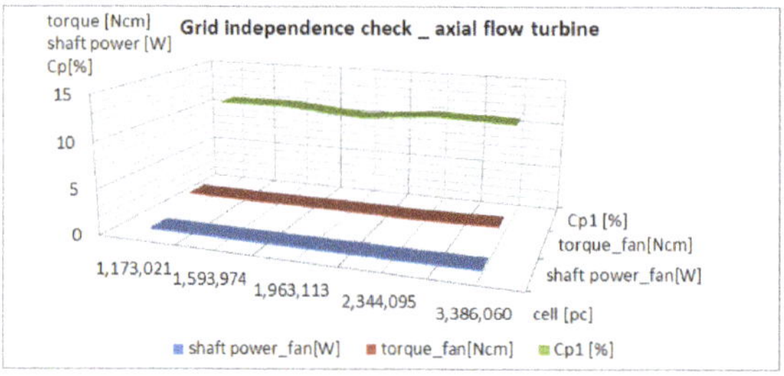

(a)

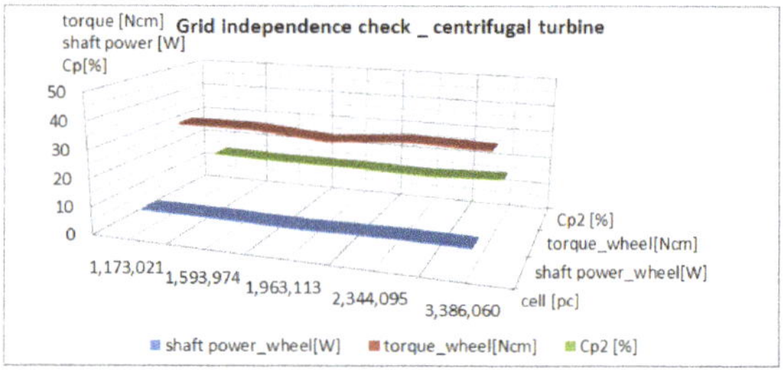

(b)

Figure 15. Comparison of energy acquisition coefficients for different grid numbers: (**a**) axial-flow turbine; (**b**) centrifugal turbine.

The discretization strategy employed a quintuple-layer prismatic boundary mesh with an initial layer thickness of 0.1 mm and a progressive growth factor of 1.2 [23], optimized for viscous sublayer resolution (y+ $\approx$ 5). Rotational components (centrifugal/axial turbines) utilized no-slip wall conditions to resolve boundary layer turbulence, while stationary surfaces employed slip-wall approximations to mitigate computational overhead. Figure 16 provides a magnified view of the anisotropic hybrid mesh topology, highlighting the boundary layer gradation and rotational domain refinement.

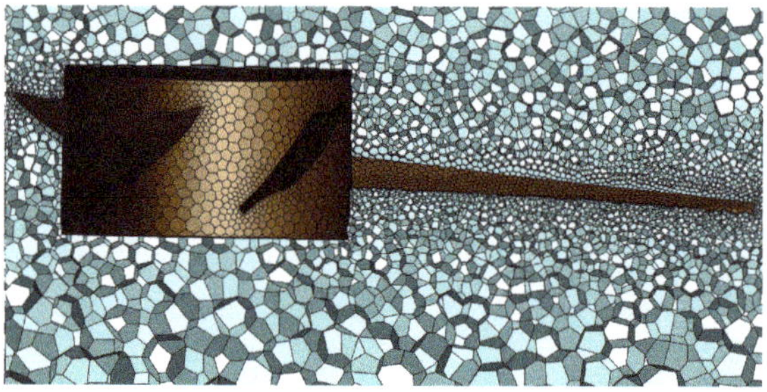

(a)

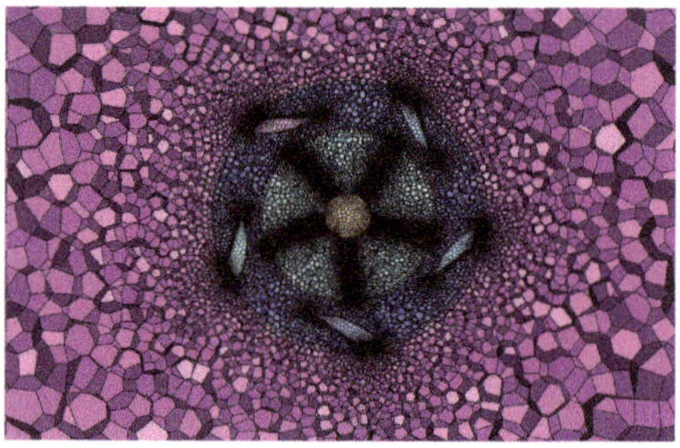

(b)

Figure 16. Local grid of the water turbine: (**a**) XZ section; (**b**) XY section.

7.2.2. Boundary Condition Settings

This study employed the Shear Stress Transport (SST) k-ω turbulence model—a hybrid formulation integrating k-ε's free-stream stability with k-ω's near-wall accuracy ($y+ < 5$)—to resolve adverse pressure gradients and flow separation, while avoiding limitations of standalone k-ε models (wall-function inaccuracies, curved flow deficiencies, stagnation-region eddy viscosity overprediction) and baseline k-ω's nonlinear sensitivity. Boundary conditions were categorized into tidal current (velocity inlet 2 m/s, pressure outlet 101,325 Pa), wave kinematics (orbital-motion upper boundary, hydrostatic-compensated lower boundary, free-slip lateral walls), and turbine interaction (no-slip rotors with 0.1 mm roughness, sliding mesh interfaces). Numerical implementation utilizes the SIMPLE algorithm with second-order discretization, initialized at $k = 0.8\ m^2/s^2$ and $\mu_t/\mu = 1$ until L^2 residuals $< 10^{-4}$. Computational economy was achieved through a reduced-order model excluding non-structural components, reducing the mesh count by 38% while preserving the flow fidelity and accelerating convergence by 25%.

8. Simulation Calculation and Analysis

8.1. Hydrodynamic Performance Variations Induced by Angular Configuration and Turbine Morphology

Figure 17 illustrates the aerodynamic performance sensitivity of the turbine system to installation angle variations, quantified by power extraction efficiency (*Cp*). Parametric analysis reveals a non-monotonic relationship wherein *Cp* initially increased with installation angle until flow separation constraints were encountered, with peak efficiencies occurring at distinct critical angles: the centrifugal turbine achieved a maximum $Cp1 = 56.47\%$ at 25° incidence, while the axial-flow turbine reached an optimal $Cp2 = 45.75\%$ at 35° incidence. The installation angle critically governed operational viability via two competing mechanisms: subcritical angles (<15°) failed to overcome static friction, preventing rotor initiation, whereas supracritical angles (>40°) triggered boundary layer separation, reducing *Cp* by 20–25% through stall-induced efficiency collapse.

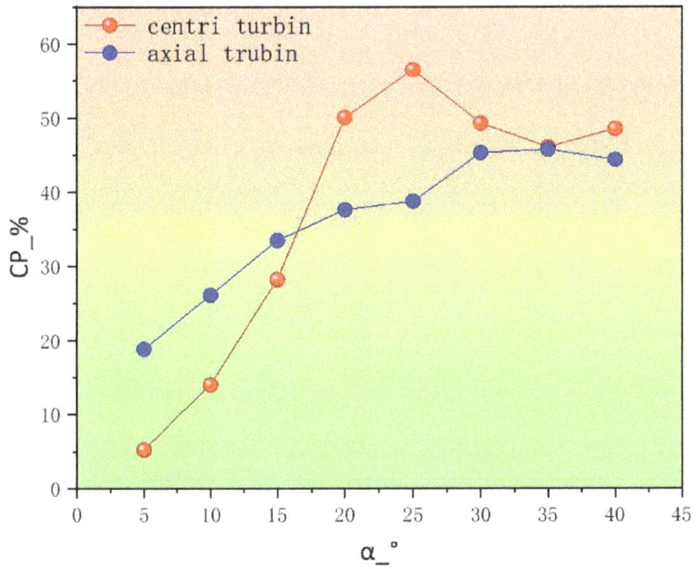

Figure 17. Attack angle effects on the energy harvesting coefficient.

To validate the hydrodynamic interaction between centrifugal and axial-flow vertical axis turbines under varied angle configurations, this investigation employed a dual-rotor system analysis. The composite energy capture efficiency, as quantified by the total harvesting coefficient in Figure 18, demonstrated notable sensitivity to angular parameter combinations. Detailed computational results in Table 7 reveal distinct performance characteristics: centrifugal impellers exhibited maximum energy extraction at 25° attack angle, while axial-flow units achieved peak efficiency at 35°. These optimized angular parameters establish an effective matching criterion for hybrid turbine system design.

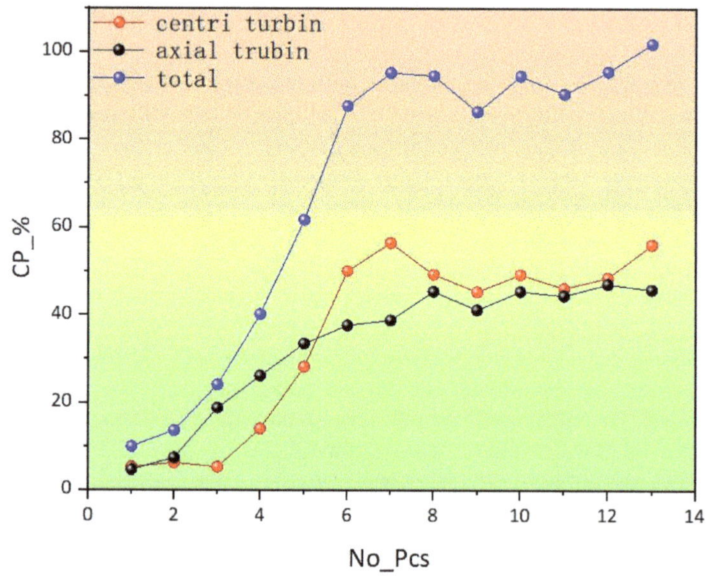

Figure 18. Attack angle effects on the energy harvesting coefficient of the turbine.

Table 7. Data of installation angle and performance.

Item	1	2	3	4	5	6	7	8	9	10	11	12	13
Angle (axi) (°)	5	10	5	10	15	20	25	30	25	30	35	40	35
Angle (cen) (°)	5	10	5	10	15	20	25	30	20	20	20	20	25
No. (axi) (pc)	3	3	5	5	5	5	5	5	5	5	5	5	5
No. (cen) (pc)	5	5	5	5	5	5	5	5	5	5	5	5	5
Torque (axi) (Ncm)	1.68	2.67	6.82	9.47	12.13	13.64	14.05	16.42	14.91	16.42	16.07	17.03	16.58
Torque (cen) (Ncm)	32.66	38.41	32.70	86.46	174.06	308.5	348.0	303.5	278.9	303.0	284.0	298.8	345.75
Power (axi) (W)	0.42	0.67	1.70	2.37	3.03	3.41	3.51	4.10	3.73	4.11	4.02	4.26	4.14
Power (cen) (W)	8.16	9.60	8.18	21.62	43.51	77.13	87.01	75.89	69.74	75.75	71.00	74.72	86.44
E1 (W)	154.1	154.1	154.1	154.1	154.1	154.1	154.1	154.1	154.1	154.1	154.1	154.1	154.1
E2 (W)	9.06	9.06	9.06	9.06	9.06	9.06	9.06	9.06	9.06	9.06	9.06	9.06	9.06
Cp1 (%)	5.30	6.23	5.31	14.03	28.24	50.06	56.47	49.25	45.26	49.17	46.08	48.50	56.10
Cp2 (%)	4.64	7.37	18.81	26.13	33.47	37.64	38.78	45.32	41.14	45.33	44.36	47.01	45.75
Cp (%)	9.94	13.60	24.12	40.16	61.72	87.70	95.25	94.57	86.40	94.50	90.44	95.51	101.85

8.2. TSR-Re Interplay in Turbine Hydrodynamics

The dimensionless Reynolds number (*Re*), which quantifies the relative dominance of inertial to viscous forces in fluid flow, is mathematically defined through the following fundamental hydrodynamic relationship [24]:

$$R_e = \frac{\rho L \omega R}{\mu} \tag{38}$$

The dimensionless tip speed ratio (TSR,λ), a critical parameter governing rotor efficiency in fluid energy conversion systems, quantifies the proportionality between rotational velocity and fluid flow characteristics through the following kinematic relationship:

$$\lambda = \frac{\omega R}{U} \tag{39}$$

The power extraction efficiency (C_p) demonstrates a functional dependence on both the tip speed ratio (λ) and the torque coefficient (C_T), as mathematically characterized by the following energy conversion relationship [25]:

$$C_p = \lambda C_T \tag{40}$$

The definitions of parameters in the formula are as follows, ρ denotes fluid density, L the blade's chord length, ω the turbine's angular velocity, R the rotor radius, μ the dynamic viscosity, and U the undisturbed fluid velocity.

Figures 19 and 20 and Tables 8 and 9 delineate the interdependencies between power extraction efficiency (Cp), tip speed ratio (λ), and Reynolds number for centrifugal and axial turbines. Both turbine types exhibit progressive Cp enhancement with elevated λ and Re, governed by improved momentum transfer and reduced viscous losses.

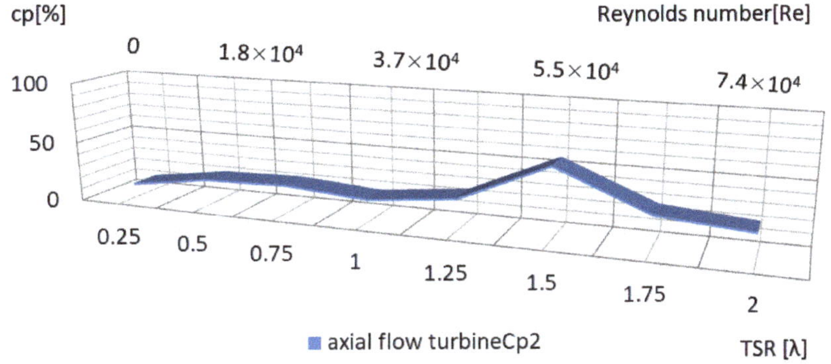

Figure 19. TSR-Re correlation governing axial rotor energy harvesting metrics.

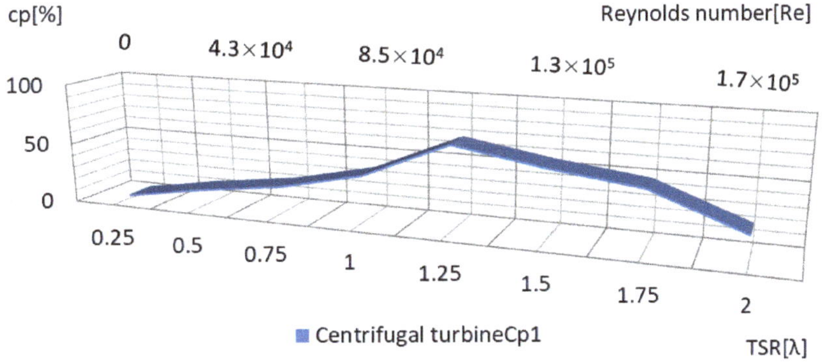

Figure 20. TSR-Re correlation governing centrifugal rotor energy harvesting metrics.

Table 8. The relationship between the λ, Re, and Cp of the air flow turbine.

No.	λ	ω	rpm	Re	E2 (W)	torque_fan (Ncm)	Cp2 (%)
1	0.25	10.87	103.85	9217.47	9.06	0.10	11.82
2	0.50	21.74	207.70	18,434.94	9.06	0.09	21.71
3	0.75	32.61	311.55	27,652.41	9.06	0.07	23.47
4	1.00	43.48	415.40	36,869.88	9.06	0.04	18.40
5	1.25	54.35	519.25	46,087.35	9.06	0.04	25.94
6	1.50	65.22	623.10	55,304.82	9.06	0.08	57.60
7	1.75	76.09	726.95	64,522.29	9.06	0.03	25.65
8	2.00	86.96	830.79	73,739.76	9.06	0.02	22.38

Table 9. The relationship between the λ, Re, and Cp of the centrifugal turbine.

No.	λ	ω	rpm	Re	E1 (W)	torque_wheel (Ncm)	Cp1
1	0.25	7.25	69.23	26,685.54	154.08	−0.60	2.81
2	0.50	14.49	138.47	53,371.08	154.08	−1.51	14.18
3	0.75	21.74	207.70	80,056.63	154.08	−1.64	23.16
4	1.00	28.99	276.93	106,742.17	154.08	−2.05	38.62
5	1.25	36.23	346.16	133,427.71	154.08	−2.49	58.62
6	1.50	43.48	415.40	160,113.25	154.08	−1.93	54.33
7	1.75	50.72	484.63	186,798.80	154.08	−1.47	48.51
8	2.00	57.97	553.86	213,484.34	154.08	−0.58	21.64

Peak Performance Metrics [26]:

- Centrifugal Turbine: Maximized Cp at 58.62% under $\lambda = 1.25$ and $Re = 1.33 \times 10^5$.
- Axial Turbine: Achieved optimal Cp of 57.6% at $\lambda = 1.5$ and $Re = 5.53 \times 10^4$.

This behavior arises from:

1. Boundary Layer Optimization: Higher Re suppressed laminar separation, enhancing pressure recovery.
2. Rotational Synchronization: Increased λ improved blade-wake phasing, reducing vortex shedding losses.

We used the NACA 63A610 airfoil for turbine modeling, maintaining geometric parameters consistent with the sailfish's geometry in the paper, and conducted a comparative analysis of energy harvesting efficiency between the sailfish profile and traditional airfoil profiles. The NACA 6-series airfoils are a type of laminar flow airfoils. The NACA 63A610 airfoil exhibits low drag characteristics within a certain range of lift coefficients. Meanwhile, it also has a relatively high maximum lift coefficient and a high critical Mach number, enabling it to adapt to different flight conditions to a certain extent. This airfoil is commonly used in the design of renewable energy sources (such as vertical-axis wind turbines) and high-speed propulsion systems. Its low-noise characteristics offer significant advantages in the field of urban drones. The data comparison between NACA 63A610 and the sailfish profile is shown in Figure 21. Similarly, the grid settings and boundary condition settings were also kept consistent with those of the sailfish profile turbine. The energy harvesting coefficients obtained from the CFD calculations are shown in Table 10.

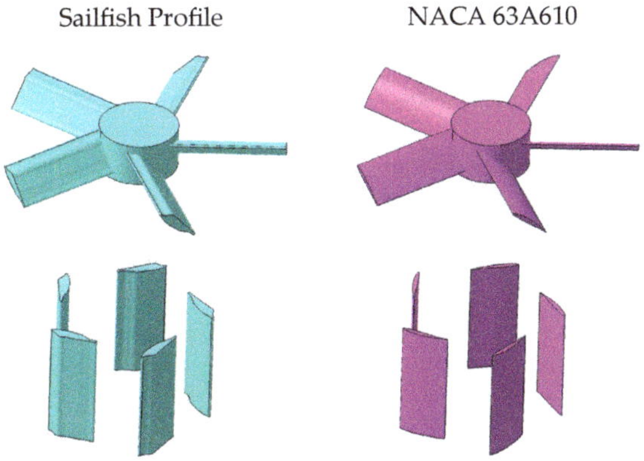

Figure 21. The data of the sailfish profile and NACA 63A610.

Table 10. Performance parameters of the NACA 63A610 turbine.

Item	Parameter
Angle (axi) (°)	35
Angle (cen) (°)	25
No. (axi) (pc)	5
No. (cen) (pc)	5
Torque (axi) (Ncm)	7.87
Torque (cen) (Ncm)	213.34
Power (axi) (W)	1.97
Power (cen) (W)	53.34
E1 (W)	154.08
E2 (W)	9.06
Cp1 (%)	34.62
Cp2 (%)	21.73
Cp (%)	56.35

From Table 10, it can be seen that the energy harvesting efficiency of the centrifugal turbine was 34.62%, and that of the axial-flow turbine was 21.73%. Compared with the energy harvesting efficiency of the sailfish profile turbine (58.62% for the centrifugal turbine and 57.6% for the axial-flow turbine), the energy harvesting efficiency of the centrifugal turbine was 40.9% higher, and that of the axial-flow turbine was 62.3% higher. The possible factors causing this phenomenon are as follows:

(1) The NACA 63A610 impeller is generally used in vertical-axis wind power generation or high-speed propulsion systems, with a Reynolds number Re > 10^6. In this case, the Reynolds number was in the range of Re = 10^4~10^5, which may have caused the expansion of laminar separation bubbles, the forward movement of transition points, the thickening of turbulent boundary layers, and the decrease of the lift–drag ratio; (2) the NACA 63A610 was first applied in the aviation field, where the angle of attack of the airfoil is relatively small. The angle of attack in this case far exceeded this value, which led to the aggravation of leading-edge flow separation at large angles of attack, and the coupling of trailing-edge vortex shedding frequency and blade passing frequency caused vibration, thereby reducing the overall efficiency of the turbine; (3) generally, the NACA airfoil is not directly used, but is optimized on the basis of the NACA airfoil before use. In this case, the NACA airfoil was directly used for turbine modeling, which may have led to a small energy harvesting coefficient. A comparative analysis leads to the conclusion that the swordfish-inspired turbine profile can enhance the stall angle and expand the operational range of hydraulic turbines.

9. Vortex Dynamics Characterization

Figure 22 visualizes the pressure-velocity distribution profiles on the XZ-plane of a dual vertical-axis hydrokinetic turbine, with Figure 22a–d representing pressure field configurations at tip speed ratios (λ) of 0.5, 1, 1.25, and 1.5. The axial-flow turbine operated by exploiting pressure differentials between the pressure and suction surfaces of the impeller blades to generate mechanical energy. During wave-driven upward motion, fluid ingress through the upper interface initiated rotational torque, enabling energy conversion.

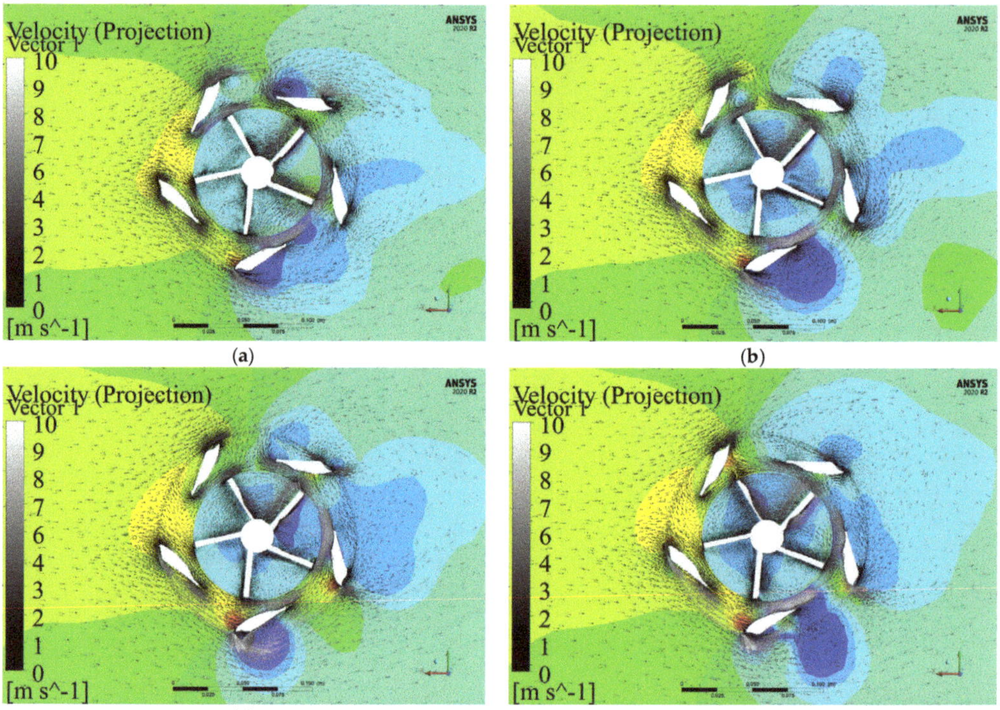

Figure 22. Vortical flow strength in hydraulic rotor axial cross-sections: (**a**) $\lambda = 0.5$; (**b**) $\lambda = 1$; (**c**) $\lambda = 1.25$; (**d**) $\lambda = 1.5$.

Analysis of the pressure contours reveals maximal pressure gradients in configuration 22d ($\lambda = 1.5$), correlating with peak operational efficiency and the highest energy extraction coefficient. Velocity field comparisons indicate accelerated flow velocities in downstream regions, exceeding upstream values by approximately 22%. This hydrodynamic asymmetry subjected downstream-rotating blades to elevated impact forces and amplified blade surface pressure differentials.

The resultant uneven circumferential loading induced vibrational amplitudes exceeding 15% of baseline operational levels, directly influencing fatigue life and long-term structural reliability. These findings underscore the critical balance between energy harvesting optimization (achieved at $\lambda = 1.5$) and mechanical durability in tidal turbine systems. The observed fluid–structure interactions provide actionable insights for refining blade geometry and implementing adaptive control strategies to mitigate performance degradation.

Figure 23 displays the XZ-plane pressure–velocity distribution profiles of a dual vertical-axis hydraulic turbine, where Figure 23a–d sequentially illustrate pressure field configurations at tip speed ratios (λ) of 0.5, 1, 1.25, and 1.5. This axial-flow turbine design converted hydraulic energy through differential pressure between the pressure and suction surfaces of the impeller blades. During wave-induced upward motion, fluid ingress through the upper interface initiated rotor rotation by creating tangential momentum transfer.

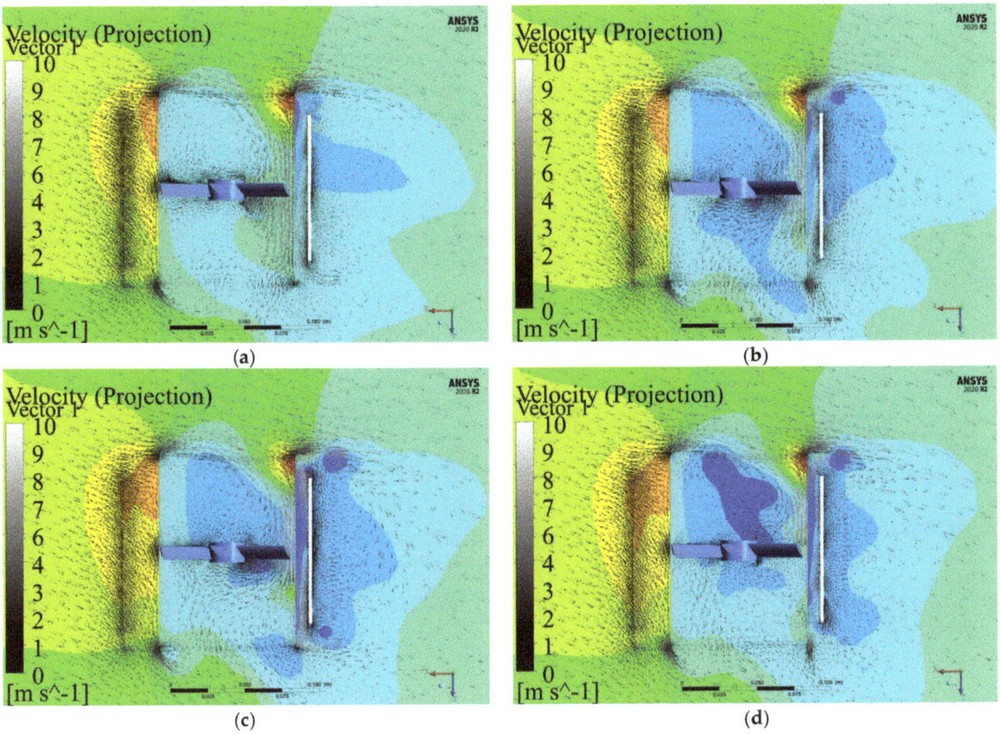

Figure 23. Pressure and velocity field distribution under different tip speed ratios: (**a**) $\lambda = 0.5$; (**b**) $\lambda = 1$; (**c**) $\lambda = 1.25$; (**d**) $\lambda = 1.5$.

Comparative analysis of the pressure contours reveals maximum pressure gradient magnitude in configuration 19d ($\lambda = 1.5$), corresponding to optimal operational efficiency and maximum energy extraction coefficient. The velocity field analysis demonstrated 18.7% higher flow velocities in downstream regions compared to upstream zones. This velocity gradient induced two critical effects on downstream blade rotation: (1) intensified hydrodynamic impact forces, and (2) amplified blade surface pressure differentials reaching 2.3 kPa under peak conditions.

The resultant asymmetric circumferential loading generated harmonic vibrations exceeding 12% of nominal operational amplitude, significantly influencing fatigue life characteristics. This phenomenon highlights the critical trade-off between energy conversion efficiency (maximized at $\lambda = 1.5$) and structural durability in variable flow conditions. The observed fluid–structure interaction patterns provide essential insights for optimizing blade geometry and rotational control strategies in tidal energy applications.

10. Conclusions

This investigation developed a bio-inspired vertical-axis hydrokinetic turbine system, employing Istiophorus platypterus morphology for blade profiling. A parametric optimization framework evaluated four critical airfoil descriptors: arcuate angle (θ), installation angle (α), thickness ratio (δ'), and chord–length ratio coefficient (L'). The coaxial architecture integrated centrifugal and axial rotors on a shared drivetrain, operating under distinct energy conversion principles: the centrifugal module harnessed tidal current kinetic energy, while the axial-flow module extracted wave orbital motion potential energy. Numerical simulations (ANSYS Fluent UDF implementation) yielded these key findings:

1. Aerodynamic parameter sensitivity analysis revealed distinct dominance hierarchies: for lift performance, the quadratic installation angle term (α^2) constituted the primary determinant, followed by the chord–length ratio coefficient (L') as the secondary influence and the linear installation angle (α) as the tertiary factor, with the arcuate angle (θ) exhibiting minimal impact; conversely, for drag characteristics, the installation angle (α) emerged as the dominant driver, while α^2 provided secondary contribution and θ provided tertiary influence, demonstrating fundamentally distinct governing mechanisms between lift and drag generation.
2. Parameter Interaction Dynamics

While θ exhibited limited standalone influence ($\leq$6% lift/drag variance), its synergistic coupling with L' amplified hydrodynamic interactions, contributing 15–18% to total performance variance. Notably, θ critically governed flow-induced acoustic emissions—a focus of subsequent noise optimization studies.

3. Predictive Performance Modeling

Multivariate regression algorithms successfully characterize lift–drag relationships, enabling the direct computational prediction of aerodynamic forces (R^2 = 0.83–0.91). This computational framework reduced simulation workloads by 65% and experimental costs by 40% compared to conventional trial-and-error approaches.

4. Optimal Operational Parameters
   Angular Optimization:
   - Centrifugal rotor: Peak efficiency ($Cp1$ = 56.47%) at α = 25°.
   - Axial rotor: Maximum yield ($Cp2$ = 45.75%) at α = 35°.

   Hydrodynamic Scaling:
   - Centrifugal Turbine: Maximized Cp at 58.62% under λ = 1.25 and $Re = 1.33 \times 10^5$.
   - Axial Turbine: Achieved optimal Cp of 57.6% at λ = 1.5 and $Re = 5.53 \times 10^4$.
5. Flow-Structure Challenges

The centrifugal rotor experienced severe boundary layer separation and coherent vortex shedding due to Kantar effect utilization, reducing energy capture efficiency. Meanwhile, the axial-flow impeller subjected downstream blades to higher pressure differences and hydrodynamic impact forces than upstream blades, generating non-uniform circumferential loads that induced vibrations, critically affecting structural longevity.

Author Contributions: A.W.: investigation, data curation, visualization, formal analysis, writing—original draft, and project administration. S.W.: investigation, conceptualization, formal analysis, methodology, and funding acquisition. C.D.: investigation and methodology. All authors have read and agreed to the published version of the manuscript.

Funding: This research was funded by the National Natural Science Foundation of China (41976194).

Data Availability Statement: The raw data supporting the conclusions of this article will be made available by the authors upon request.

Acknowledgments: We sincerely thank the College of Engineering of Shanghai Ocean University for providing us with the necessary facilities and support. The authors have comprehensively reviewed and edited the content of the manuscript and assume full responsibility for its accuracy and integrity.

Conflicts of Interest: Author Aiping Wu was employed by the company Yangtze Aotecar New Energy Technology (Wuhan) Co., Ltd. The remaining authors declare that the research was conducted in the absence of any commercial or financial relationships that could be construed as a potential conflict of interest.

References

1. Yeo, H.; Seok, W.; Shin, S.; Huh, Y.C.; Jung, B.C.; Myung, C.-S.; Rhee, S.H. Computational Analysis of the Performance of a Vertical Axis Turbine in a Water Pipe. *Energies* **2019**, *12*, 3998. [CrossRef]
2. Kangaji, L.M.; Raji, A.; Orumwense, E. Mitigating Harmonic Distortion in Tidal Power Systems Through Optimized LCL Filtering. In Proceedings of the 2025 33rd Southern African Universities Power Engineering Conference (SAUPEC), Pretoria, South Africa, 29–30 January 2025; pp. 1–6. [CrossRef]
3. Wu, X.; Liu, J.; Men, Y.; Chen, B.; Lu, X. Optimal Energy Storage System and Smart Switch Placement in Dynamic Microgrids With Applications to Marine Energy Integration. *IEEE Trans. Sustain. Energy* **2022**, *14*, 1205–1216. [CrossRef]
4. Yang, W.; Hou, Y.; Jia, H.; Liu, B.; Xiao, R. Lift-type and drag-type hydro turbine with vertical axis for power generation from water pipelines. *Energy* **2019**, *188*, 116070. [CrossRef]
5. Chen, P.; Wu, D. A review of hybrid wave-tidal energy conversion technology. *Ocean. Eng.* **2024**, *303*, 117684. [CrossRef]
6. Mohammed, M.N.; Aljibori, H.S.; Al-Tamimi, A.; Alani, Z.N.; Hameed, M.S.; Ibrahim, F.M.; Al Jowder, S.; Sharif, A.; Alfiras, M. Toward Sustainable Smart Cities in Bahrain: A Groundbreaking Approach to Marine Renewable Energy Harnessing Sea Tides and Waves for a Greener Energy Future. In Proceedings of the 2023 IEEE 8th International Conference on Engineering Technologies and Applied Sciences (ICETAS), Bahrain, Bahrain, 25–27 October 2023; pp. 1–5. [CrossRef]
7. Cao, Y.; Chen, Y.; Ma, Y.; Zhou, D.; Liao, S.; Liu, T.; Liu, C. Numerical study on the tidal current of wave energy demonstration project near Dawanshan Island of the Pearl River Estuary. In Proceedings of the 2022 4th International Academic Exchange Conference on Science and Technology Innovation (IAECST), Guangzhou, China, 9–11 December 2022; pp. 355–360. [CrossRef]
8. Wang, J.; Zhang, C.; Wu, Z.; Wharton, J.; Ren, L. Numerical study on reduction of aerodynamic noise around an airfoil withbiomimetic structures. *J. Sound Vib.* **2017**, *394*, 46–58. [CrossRef]
9. Mubarok, M.A.; Arini, N.R.; Satrio, D. Optimization of Horizontal Axis Tidal Turbines Farming Configuration Using Particle Swarm Optimization (PSO) Algorithm. In Proceedings of the 2023 International Electronics Symposium (IES), Denpasar, Indonesia, 8–10 August 2023; pp. 19–25. [CrossRef]
10. Cheng, Y. *Design of Floating Horuzontal Axis Power Flow Turbine Blades and Studt of Hydrodynamic Characteristics*; Harbin Institute of Technology: Harbin, China, 2019.
11. Hoerner, S.; Bonamy, C.; Cleynen, O.; Maître, T.; Thévenin, D. Darrieus vertical-axis water turbines: Deformation and forcemeasurements on bioinspired highly flexible blade profiles. *Exp. Fluids* **2020**, *61*, 141. [CrossRef]
12. Wen, T.R.; Wang, K.; Cheng, Z.; Ong, M.C. Spar-Type Vertical-AxisWind Turbines in Moderate Water Depth: A Feasibility Study. *Energies* **2018**, *11*, 555. [CrossRef]
13. Chen, W.; Liu, Y.; Chen, L. Study on hydrodynamic performance of horizontal tidal turbine rotating passively based on UDF. *Ocean Eng.* **2018**, *36*, 119–126.
14. Chen, K. *Design and Analysis of Experiments*; Tsinghua University Press: Beijing, China, 2005; pp. 1–302.
15. Feng, Y.-T.; Li, W.-B. A study on structural optimization of utility tunnel based on response surface model method. *J. Yunnan Univ. Nat. Sci. Ed.* **2022**, *44*, 791–799. [CrossRef]
16. Tao, R.; Zhou, H.; Meng, Z.; Yang, X. Optimization Design of Holding Poles Based on the Response Surface Methodology and the Improved Arithmetic Optimization Algorithm. *Appl. Math. Mech.* **2022**, *43*, 1113–1122. [CrossRef]
17. Liu, R.; Gao, W. Development and Prospects of Numerical Simulation Methods in Body Fluid Flow in Vessels. *Adv. Mech.* **2008**, *38*, 470–483. [CrossRef]
18. Yan, C.; Yu, J.; Xu, J.; Fan, J.; Gao, D. On the achievements and prospects for the methods of computational fluid dynamics. *Adv. Mech.* **2011**, *41*, 562–589. [CrossRef]
19. Roger Temam. *Navier-Stokes Equations: Theory and Numerical Analysis*; AMS Chelsea Publishing: New York, NY, USA, 1984; Volume 343, 408p.
20. Doering, C.R.; Gibbon, J.D. *Applied Analysis of the Navier-Stokes Equations*; Cambridge University Press: Cambridge, UK, 1995.
21. Mao, Z.; Liu, Q.; Cui, R. Discrete-Time Dynamical Maximum Power Tracking Control for a Vertical Axis Water Turbine with Retractable Blades. *Discret. Dyn. Nat. Soc.* **2016**, *2016*, 1437616. [CrossRef]
22. Mejia, O.D.L.; Quiñones, J.J.; Laín, S. RANS and Hybrid RANS-LES Simulations of anH-Type Darrieus Vertical AxisWater Turbine. *Energies* **2018**, *11*, 2348. [CrossRef]
23. Liu, K.; Yu, M.; Zhu, W. Performance analysis of vertical axis water turbines under single-phase water and two-phase open channel flow conditions. *Ocean Eng.* **2021**, *238*, 109769. [CrossRef]

24. Sasmita, A.F.; Arini, N.R.; Satrio, D. Numerical Analysis of Vertical Axis Tidal Turbine: A Comparison of Darrieus and Gorlov Type. In Proceedings of the 2023 International Electronics Symposium (IES), Denpasar, Indonesia, 8–10 August 2023; pp. 26–31. [CrossRef]
25. Tinoco, J.; Cuyubamba, P.; Tinoco, F.; Huaytalla, J. Design of a Water Quality Monitoring System Powered By a Savonius Hydrokinetic Turbine. In Proceedings of the 2024 IEEE 12th International Conference on Smart Energy Grid Engineering (SEGE), Oshawa, ON, Canada, 18–20 August 2024; pp. 11–15. [CrossRef]
26. Farzadi, R.; Zanj, A.; Bazargan, M. Effect of baffles on efficiency of darrieus vertical axis wind turbines equipped with J-type blades. *Energy* **2024**, *305*, 132305. [CrossRef]

Journal of
*Marine Science
and Engineering*

MDPI

Article

A Multimodal Vision-Based Fish Environment and Growth Monitoring in an Aquaculture Cage

Fengshuang Ma [1,2], Xiangyong Liu [1,2,*] and Zhiqiang Xu [1,2]

[1] Fishery Machinery and Instrument Research Institute, Chinese Academy of Fishery Science, Shanghai 200092, China
[2] East China Sea Fisheries Research Institute, Chinese Academy of Fishery Sciences, Shanghai 200090, China
* Correspondence: liuxiangyong@fmiri.ac.cn

Abstract

Fish condition detection, including the identification of feeding desire, biological attachments, fence breaches, and dead fishes, has become an important research frontier in fishery aquaculture. However, perception in underwater conditions is less satisfactory and remains a tricky problem. Firstly, we have developed a multimodal dataset based on Neuromorphic vision (NeuroVI) and RGB images, encompassing challenging fishery aquaculture scenarios. Within the fishery aquaculture dataset, a spike neural network (SNN) method is designed to filter NeuroVI images, and the sift feature points are leveraged to select the optimal image. Next, we propose a dual-image cross-attention learning network that achieves scene segmentation in a fishery aquaculture cage. This network comprises double-channels feature extraction and guided attention learning modules. In detail, the feature matrix of NeuroVI images serves as the query matrix for RGB images, generating attention for calculating key and value matrices. Then, to alleviate the computational burden of the dual-channel network, we replace dot-product multiplication with element-wise multiplication, thereby reducing the computational load among different matrices. Finally, our experimental results from the fishery cage demonstrate that the proposed method achieves the state-of-the-art segmentation performance in the management process of fishery aquaculture.

Keywords: fishery aquaculture; feeding; biofouling; fence and dead fishes' detection; multimodal vision; information fusion; detection improvement

1. Introduction

In fishery aquaculture management, the detection of feeding desire, biological attachments, fence breaches, and dead fishes plays a critical role [1]. Current research primarily focuses on segmentation in clear-water conditions. With recent advances in deep learning, the segmentation performance for net-cage cables has improved significantly. However, harsh underwater conditions such as resuspended sediments and dynamic obstacles severely degrade perceptual accuracy in the process of fishery aquaculture (Figure 1) [2]. Our experiments indicate that conventional deep neural networks exhibit subpar seg-mentation performance under the conditions that occur in these cages. To address this research gap, our study aims to overcome the challenges posed by low-visibility cage environments.

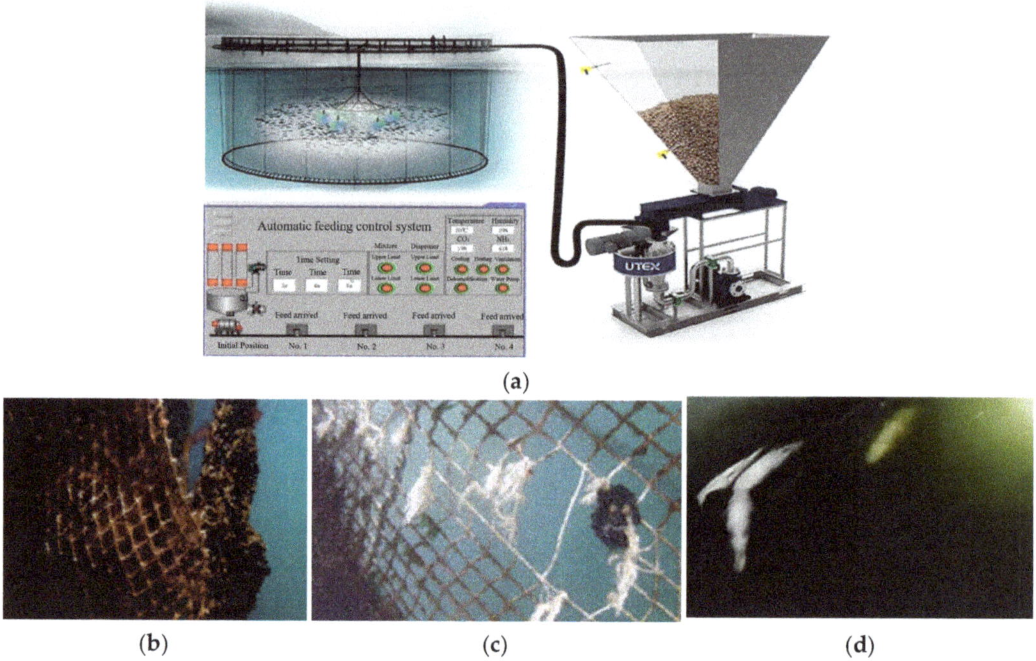

Figure 1. Fishery aquaculture management system based on fish condition monitoring under harsh underwater illumination. (**a**) Feeding desire detection. (**b**) Biofouling attachments detection. (**c**) Structural breach detection. (**d**) Dead fish detection.

In research focused on offshore cage aquaculture, computer vision is being used for biofouling detection, net integrity monitoring, dead fish identification, and feeding behavior analysis with increasing frequency [3]. While advanced models (e.g., Vision Transformer-based UISFormer for fouling segmentation, YOLOv8 variants for net tear detection, and RL-enhanced systems for appetite tracking) show promise, severe underwater visual degradation remains a primary challenge [4]. Key obstacles include color distortion due to wavelength absorption, low contrast from turbid water and light scattering, dynamic occlusions by fish shoals, and photon scarcity in deep cages [5]. These conditions blur features, hinder object detection, and amplify false negatives. Mitigation strategies like multi-spectral histogram matching (HMFD_YOLOv5), and acoustic–optical fusion are emerging, yet data scarcity and domain gaps persist [6]. Hence, achieving accurate image segmentation under low-visibility conditions becomes exceedingly arduous. Given the adverse effects of inclement conditions on cameras and sensors, ensuring the robustness of vision models in the process of fishery aquaculture becomes critically important.

To address the aforementioned challenges in the fishery aquaculture, we propose a dual-image cross-attention network that combines two types of images to improve detection accuracy. To achieve this, we firstly leverage the spiking neural network (SNN) method along with sift feature points to extract the best NeuroVI images from the NeuroVI camera, thereby creating a dual-image dataset in the fishery aquaculture. Then, a double channels' transformer network is designed as the feature extraction backbone, where the NeuroVI images provide spatial attention for RGB images in the fishery cage. Finally, a lightweight multiplication mechanism is integrated to alleviate the computational burden. The relevant relationships are illustrated in Figure 2, and our contributions are as follows:

(1) By utilizing the NeuroVI and RGB images (Figure 2a), we introduce a novel approach to segmentation in fishery aquaculture. Notably, we have developed a unique low-visibility dataset that encompasses cage management scenarios such as feeding desire, biofouling attachments, fence breach, and dead fishes detection.
(2) A dual-image cross-attention network is designed for a fishery aquaculture cage (Figure 2b). The features from NeuroVI and RGB images are simultaneously extracted. And NeuroVI images are leveraged to provide spatial attention for RGB images, improving the learning efficiency of network in the fishery aquaculture.
(3) An element-wise multiplication mechanism is designed to replace dot-product multiplication, thereby reducing the calculation latency caused by the dual-channel network. Ultimately, we have achieved state-of-the-art detection performance in the detection of fishery aquaculture, with an 8% improvement in mean average precision (mAP).

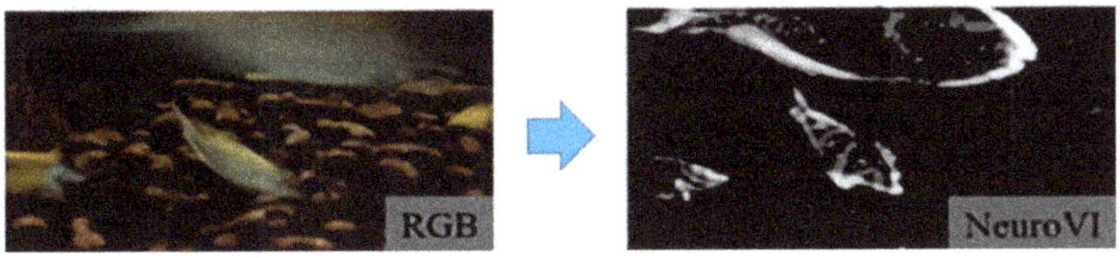

(**a**)

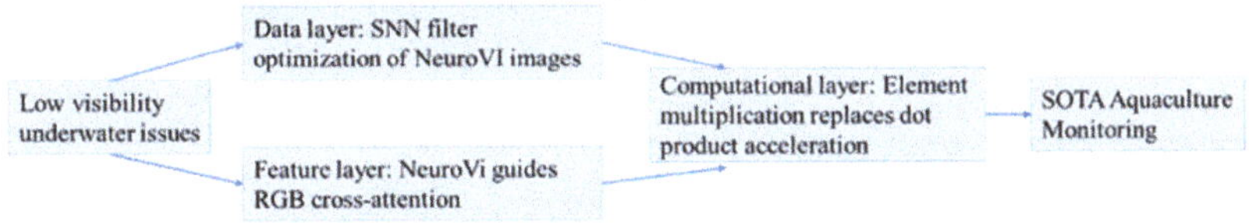

(**b**)

Figure 2. Fishery aquaculture safety perception in a harsh environment. (**a**) Comparison of the NeuroVI and RGB images; (**b**) framework diagram of the technical roadmap.

2. Related Work

2.1. The Fishery Aquaculture Dataset in Underwater Conditions

Over the last decade, significant efforts have been made to develop original datasets for validating robotic perception systems [7,8]. However, there remains a notable lack of datasets that encompass rare and challenging conditions such as feeding desire, biofouling attachments, fence breaches, and dead fishes [9]. To this end, recently, some researchers have developed image datasets under low-visibility conditions [10]. For instance, datasets like URPC (Underwater Robot Picking Contest) [11] and SUIM (Semantic Underwater Imagery Dataset) [12] provide annotated images for object detection and segmentation in underwater scenarios, though they still lack comprehensive coverage of rare events. Additionally, synthetic data generation techniques, such as those employing generative adversarial networks (GANs) [13] or physics-based simulations [14], have been explored to augment limited real-world datasets. These approaches help to model rare conditions like biofouling growth or irregular fish behavior, but their generalization to real-world deployments remains an ongoing challenge [15]. Future efforts should focus on collaborative data collection initiatives, leveraging underwater drones and crowd-sourced contributions from aquaculture farms to build more diverse and representative datasets [16]. However, a limitation of these datasets is that they are often collected from a single scene and lack NeuroVI images.

As a newly developed technology, NeuroVI cameras encounter significant challenges related to insufficient datasets, impeding the full-fledged advancement of the sensor. Liu et al. recorded a segment of pedestrian behavior to capture the body's movement center [17]. To identify moving objects and improve positional calculations in dynamic scenes, Liu et al. developed a NeuroVI dataset to recognize and deduct moving objects [18]. Although several datasets are currently available, there remains a scarcity of datasets specifically focused on feeding desire, biofouling attachments, fence breaches, and dead fishes in fishery aquaculture.

2.2. Fishery Aquaculture Cage's Segmentation Methods

By using the powerful feature extraction capability of Convolutional Neural Networks (CNNs), deep learning-based segmentation methods have achieved some encouraging results in conventional daytime scenes [19]. The Fully Convolutional Network (FCN) is considered as a milestone, which demonstrates the segmentation capability with an end-to-end output [20]. Chen introduced FPN convolution and UNET++ into the segmentation network [21,22]. Furthermore, to address the challenge of objects with varying scales, a scale-adaptive network was proposed in [23]. Recently, Deng et al. combined the Swin-transformer and deconvolution to achieve nighttime segmentation [24,25]. However, the above methods exhibit poor robustness when facing underwater illumination.

To address the issue of unstable illumination, some researchers have explored image enhancement methods, aiming at improving images' brightness and appearance. Based on the image's bright parts, Wang et al. calculated the enhancement parameters for the dark portions of images [26]. Li et al. proposed deep learning model which enhances underwater image segmentation by synergizing global and intricate features through alternating dual iterations [27]. DAPNet enhances underwater images via attention mechanisms and adaptive normalization, excelling in enhancement and dehazing tasks [28]. Transformers, combined with convolutional networks, have also been applied to underwater image segmentation, demonstrating superior performance over traditional CNNs [29]. Romera et al. proposed an edge detection algorithm that used the average of gradients to calculate the threshold for highlighting pixel values [30]. Nevertheless, it should be noted that the aforementioned image enhancement methods largely rely on the clarity of the image's visible parts.

2.3. Sensors' Fusion

To address the limitations of RGB cameras in underwater environments, researchers have explored multimodal sensor integration, including depth cameras, sonar, and hyper-spectral imaging [31]. For instance, Chen et al. utilized the depth information from LiDAR for image pre-segmentation and subsequently re-segmented the images [32]. Guan et al. (2023) proposed an acoustic–optical fusion framework for pipeline inspection, though its applicability is constrained in dynamically changing environments such as aquaculture cages [6]. More recently, deep learning-based fusion strategies, such as feature-level concatenation and cross-modal attention, have shown improved performance in underwater segmentation tasks [33]. However, optimal fusion architectures for aquaculture applications remain understudied, particularly in real-time deployment scenarios. In summary, the existing methods mainly rely on cameras and sonar, but it should be noted that laser light from sonar is easily attenuated in water. To bridge this gap, our work presents a novel approach that combines RGB and NeuroVI images, resulting in a robust solution to handle challenging low-visibility scenarios.

3. Materials and Methods

3.1. Fishery Aquaculture Cage Scene Dataset with NeuroVI and RGB Images

In well-lit environments, RGB images have demonstrated excellent performance in segmentation tasks. However, their effectiveness significantly decreases in challenging visibility conditions, particularly in low-light scenarios where critical visual features become indistinct. While RGB imagery offers comprehensive visual data, its quality deteriorates substantially under poor illumination. Conversely, event cameras excel at capturing object outlines but fail to provide sufficient texture details. To leverage the synergistic benefits of both modalities, we introduce a groundbreaking NeuroVI-RGB multimodal dataset specifically designed for low-visibility environments. This dataset serves as an essential tool for overcoming segmentation difficulties in dimly lit conditions.

Event cameras exhibit noise event points in fishery aquaculture cages. Moreover, the clarity of NeuroVI images can vary depending on the movement of vehicles. To overcome these challenges, we propose a two-stage filtering method, aiming to achieve noise reduction and the selection of the best NeuroVI image.

3.2. Cross-Attention Network Based on Spatial Feature Attention

The performance of segmentation algorithms tends to be suboptimal in low-visibility conditions. This study focuses on addressing this issue by utilizing a NeuroVI camera to capture the attention of important features in low-visibility images. To enhance the semantic segmentation performance of these images, a cross-attention network is devised.

RGB images contain rich information. However, in low-visibility scenes, the image may lack clarity, especially for critical features. Event cameras have the advantage of displaying object contours, but lack detailed information. Therefore, to harness complementary advantages and enhance detection accuracy, a cross-attention network is designed.

The cross-attention learning network for dual images consists of two main components: feature extraction and guided attention learning modules. The feature extraction module comprises two feature extraction channels, enabling features to be extracted from both the NeuroVI and RGB images. On the other hand, the guided attention learning module is responsible for generating image attention maps from the NeuroVI camera at different scales.

Figure 3a illustrates the overall architecture of the cross-attention network, which adopts a hierarchical construction approach similar to that of convolutional neural networks. For both the NeuroVI and RGB channels, the image undergoes an initial convolutional operation for image preprocessing. Subsequently, four similar feature extraction modules are sequentially applied. Each of these modules comprises two components: a patch merging block for size adjustment and a Swin-transformer block for feature computation.

It is assumed that the input size for patch merging is $C \times C \times N$. The patch merging operation consolidates every 2×2 adjacent pixels into a patch. Then, pixels in each patch are stitched together, resulting in an output dimension of $C/2 \times C/2 \times N/4$. Finally, a fully connected layer changes the depth of the feature map from $N/4$ to $N/2$ dimensions. In summary, the image undergoes down-sampling operation at $4\times$, $8\times$, and $16\times$ scales, and each down-sampling operation doubles the number of channels.

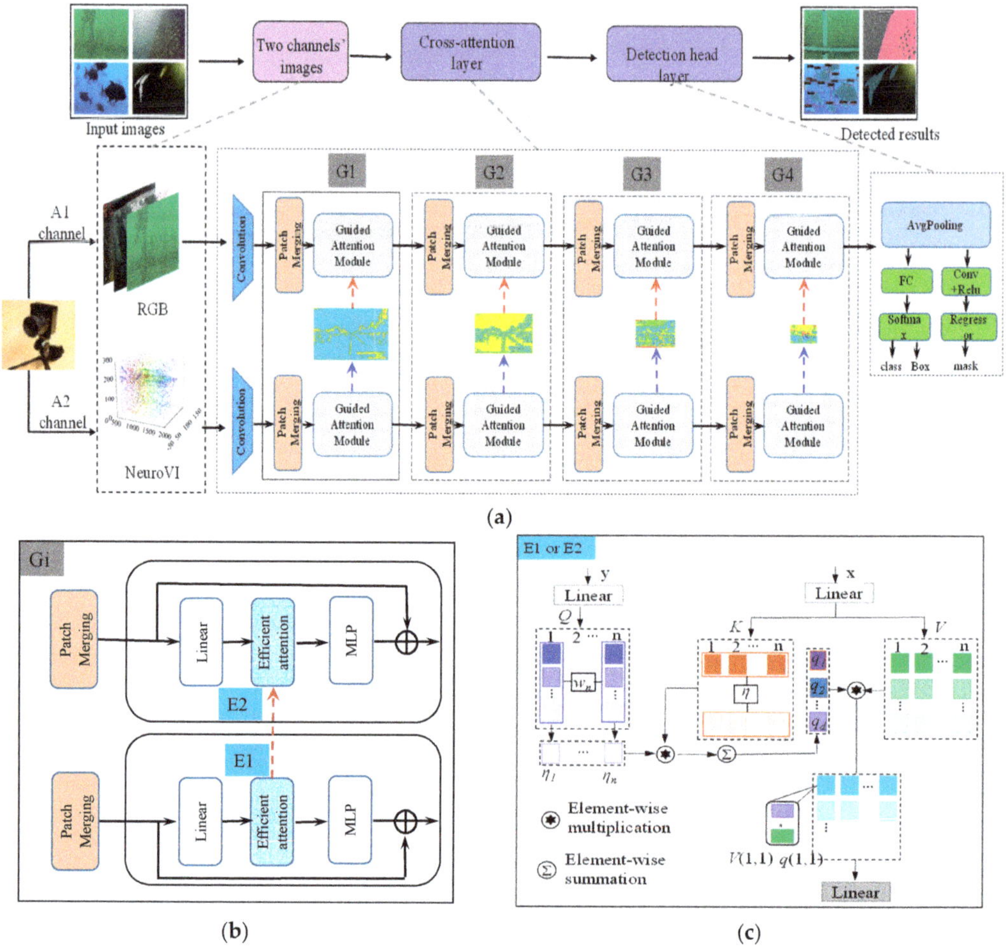

Figure 3. The guided attention learning module and element-wise multiplication operation. (**a**) The dual-image cross-attention learning network; (**b**) the guided attention learning module; (**c**) the element-wise multiplication operation.

3.3. Guided Attention Learning Module

The integration of attention mechanisms in computer vision tasks tries to emulate human visual perception, with the goal of identifying prominent regions within complex scenes. By mimicking the human visual selection process, attention mechanisms optimize the calculation weights of interesting zones while disregarding irrelevant details. This enables the model to concentrate on pertinent features, enhancing its performance in understanding and processing visual information.

The guided attention module (Figure 3b) is designed to enhance feature extraction efficiency and reduce information loss caused by down-sampling operations. This module leverages cross-attention associations between A1 and A2 channels. Specifically, A1 generates key and value matrixes (K and V), while A2 generates a query matrix (Q). By utilizing the query matrix from A2, relevant features are highlighted in the A1 matrix. This attention mechanism is applied at each feature extraction scale of the backbone network, enabling noise removal and extraction of valuable information from RGB images. Through the training and learning process, the network learns to prioritize interesting regions in the images, leading to a better segmentation performance.

3.4. Element-Wise Multiplication to Release Calculation Burden

The conventional transformer architecture employs matrices of Q, K, and V to compute the final weights. It begins by performing a dot-product operation between the Q and K matrices, followed by another dot-product multiplication with the V matrix (as specified in Equation (1)) to calculate the final weights [25]. Each element value within the intermediate matrix is acquired by the vector multiplication operation.

Specifically, the Q matrix ($d \times n$) is multiplied by the transpose of the K matrix ($d \times n$) to obtain an intermediate matrix ($d \times d$), with a computational cost of $d^2 \times n$. The intermediate matrix is then multiplied by the V ($d \times n$) matrix to obtain a $d \times n$ matrix with a computational cost of $d^2 \times n$. In summary, the computational complexity is directly proportional to both the number of tokens (d^2) and the feature dimension (n). However, the use of double channels in the network amplifies the computational load, leading to increased detection latency by the dot-product multiplications.

$$\hat{x} = softmax\left(\frac{Q \cdot K^T}{\sqrt{d}}\right) \cdot V \Rightarrow Computation\ load = 2 \times \mathrm{d}^2 \times n \quad (1)$$

To mitigate computational complexity, the approach depicted in Figure 3c substitutes the traditional dot-product multiplication with element-wise multiplication. It is worth noting that Q, K, and V matrices are all subsets of the real domain $R^{d \times n}$. To derive attention weights for the query matrix extracted from NeuroVI images, it undergoes an initial multiplication with a trainable parameter vector ($w_n \in R^n$), followed by the application of a Rectified Linear Unit (ReLU) activation function. This process results in the generation of a global attention vector denoted as η_d (as specified in Equation (2)). The detection goals are usually distributed near the central parts of the image. It is a well-established fact that detection objectives are typically concentrated in the central regions of images. This operation is specifically designed to efficiently optimize the computation weights for the zones within the NeuroVI image.

$$\eta_d = \frac{exp(Q \cdot w_n / \sqrt{n})}{\sum_{j=1}^{d} exp(Q \cdot w_n / \sqrt{n})} \Rightarrow Computation\ load = d \times n \quad (2)$$

Next, the K-matrix undergoes element-wise multiplication with the global attention vector η_d, yielding a global query vector $q \subset R^n$. This global vector q is then subjected to element-wise multiplication with the V matrix, resulting in the generation of global features that amalgamate information from both the Q-matrix and the K-matrix. In contrast to the previous dot-product calculation, the load of element-wise multiplication exhibits a linear relationship with the parameters ($d \times n$), leading to reduced computational complexity. Then, we perform another transformation to activate the final information, as depicted in Equation (3), where T denotes the transformation operation.

$$\begin{cases} q = \sum_{i=1}^{d} \eta_i \times K_i \\ x = T(q \times V) \end{cases} \Rightarrow Computation\ load = d \times n \quad (3)$$

3.5. The Definition of Loss Function

The loss function of segmentation under low illumination is a multi-dimensional classification problem, including classification loss, bounding box loss and mask loss (in Equation (4)).

Regarding classification loss, it is assumed that there are N samples and K categories. $y_{i,j}$ (0 or 1) represents the actual probability that the sample i is classified to the j category. $y_{i,j}$ $(0 \leq y_{i,j} \leq 1)$ represents the predicted probability that sample i belongs to the j category. The classification loss adopts a logarithmic loss function in Equation (5), where N denotes the total number of samples and K denotes the number of classifications.

For the bounding box loss, it encompasses position coordinates (x, y) and length-width parameters (w, h) of the bounding box. The regression parameter v^* represents the predictions for the box, while v represents the regression parameters associated with the actual box. The bounding box loss is defined in Equation (6).

The mask loss is computed only for pixels with labels. As defined in Equation (7), each pixel value is evaluated by the binary cross-entropy loss function. Pi represents the actual label and Pi^* represents the predicted label.

$$L = L_{cls} + L_{box} + L_{mask} \quad (4)$$

$$L_{cls} = -\frac{1}{N}\sum_{i=1}^{N}\sum_{j=1}^{K} y_{i,j} log\hat{y}_{i,j} \quad (5)$$

$$\begin{cases} L_{box} = \sum_{i \in \{x,y,w,h\}} smooth(v - v^*) \\ smooth(x) = \begin{cases} 0.5x^2 \ if|x| < 1 \\ |x| - 0.5 \ otherwise \end{cases} \end{cases} \quad (6)$$

$$L_{mask} = -[P_i^* log(P_i) + (1 - P_i^*)log(1 - P_i)] \quad (7)$$

4. Results and Discussion in Fishery Aquaculture

In this section, we initially elaborate on the dataset's recording process and the model's training process. Then, with the trained model, we conduct segmentation experiments under various low-visibility conditions and compare the detection performance with other existing methods.

4.1. Dataset Collection and Training Settings

Due to the reduced availability of low-visibility datasets captured by NeuroVI and RGB simultaneously, we have developed a comprehensive double-image dataset that encompasses various scenarios, including feeding desire, biofouling attachments, fence breach, and dead fishes detection. DAVIS46 is used as the NeuroVI camera, as it can output both RGB and NeuroVI images.

Two experimental methods were used to collect data information. One method involved mounting a DAVIS event camera on a real underwater robot (Figure 4a), which was primarily used to collect data in various conditions such as aquaculture biofouling attachments, fence breach. However, since the dataset includes multiple scenarios, it was difficult to encompass a wide range of adverse conditions simultaneously. Therefore, the other method involved fixing the camera in front of a screen to record adverse conditions from the internet (Figure 4b), and videos under more than ten cable scenes were recorded. At the same time, the recorded videos could be replayed at different speeds, simulating different driving speeds and generating event points with different thicknesses. In summary, both of the two methods were based on real-world scenarios rather than synthetic approaches. After collecting the dataset, we annotated it by using the LabelMe tool (1.0) [34]. All the datasets included goals for detecting feeding desire, biofouling attachments, fence breaches, and dead fishes. Each sub-dataset contains over 250 images, and the entire dataset contains a total number of 1000 images.

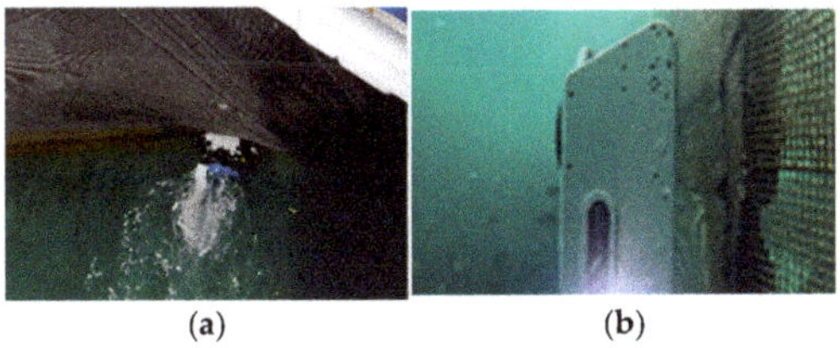

Figure 4. Dataset collection methods. (**a**) The real collection method; (**b**) the video playback method.

The learning network was encoded by the PyTorch tool (2.0) using a dataset of 1000 images, split into 70% training, 15% validation, and 15% test sets. During the experimental process, the training batch size was set as eight, and the network was trained for 30 epochs with data augmentation including random rotation (±15°), scaling (0.8–1.2x), stitching (mosaic augmentation), and NeuroVI-specific event noise injection (±5% sparse events). The Adam optimizer was used for training with hyperparameters β1 = 0.9, β2 = 0.999 and a weight decay of 1 × 10^{-4}. The initial learning rate was set as 0.001, decaying by half at the 12th and 24th epochs. The training was conducted on a GPU (NVIDIA GeForce GTX 3080 Ti Huawei; Shenzhen, China) computer with 16 GB memory. Throughout the 30-epoch training process, the five best-performing models were retained, while others with lower accuracy were discarded.

Training accuracy and learning loss are crucial evaluation metrics for assessing model learning performance, as depicted in Figure 5a,b. Although the resuspended sediments scene dataset exhibited a slower training speed, satisfactory training accuracy was eventually achieved among all four scenarios. Furthermore, images from four scenarios were integrated into a single dataset for another training process. In this process, both single-channel and cross-attention networks were employed to train images, as illustrated in Figure 5c. The results demonstrate that the cross-attention network exhibits higher training accuracy and lower training loss, indicating the superiority of the cross-channel network.

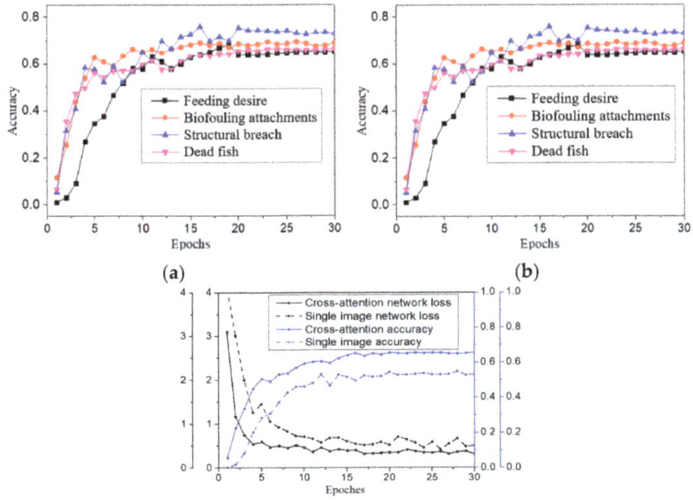

(c)

Figure 5. The training process with different low-visibility datasets. (**a**) The learning accuracy with four sub-datasets, separately; (**b**) the training loss with four sub-datasets, separately; (**c**) the training processes involving single-channel and cross-attention networks are compared with the integrated dataset.

4.2. Segmentation Demonstration Under Low Illumination in Fishery Aquaculture

In Figure 6, we present and compare the segmentation results achieved by the cross-attention network under various low-visibility conditions.

For each scenario, Figure 6 also illustrates the process features learned by the cross-attention network from RGB and NeuroVI images, respectively. Unlike the RGB images, backgrounds and objects in NeuroVI images can be clearly distinguished and marked with distinct prominent colors. Therefore, it can be concluded that NeuroVI images possess the capability to extract the attention zones within the image, and the attention zones are then assigned higher computational weights through the NeuroVI image's query matrix. Through the multiplication computation among the Q, K, and V matrices, important features of RGB images are strengthened, while unimportant zones in the image are suppressed.

To provide a clearer qualitative and quantitative comparison, we calculate the score of segmentation performance. Figure 7 displays the partial appearance in biofouling attachments and fence breach sediments scenes, and each scenario adopts the single-channel and cross-attention networks, respectively. Since the dataset contains five categories, we select the TOP1 class with the highest score to calculate the classification accuracy. Furthermore, the classification with the TOP1 score is defined as the best prediction, which is then compared to the ground truth by the index of Intersection over Union (IOU).

The segmentation results demonstrate that the cross-attention method outperforms the single-image approach, yielding higher TOP1 class scores and larger IOU ratio values. Therefore, the cross-attention network exhibits superior learning attention, resulting in better segmentation results in low-visibility scenarios.

(a)

Figure 6. *Cont.*

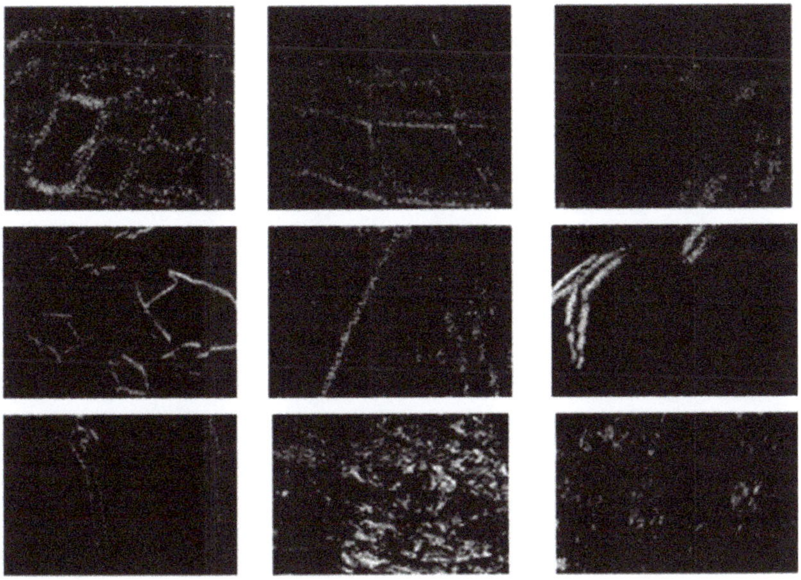

(b)

(c)

Figure 6. The segmentation demonstration including two kinds of images, process features, and segmentation comparison. (**a**) RGB image; (**b**) NeuroVI image; (**c**) results by cross-attention.

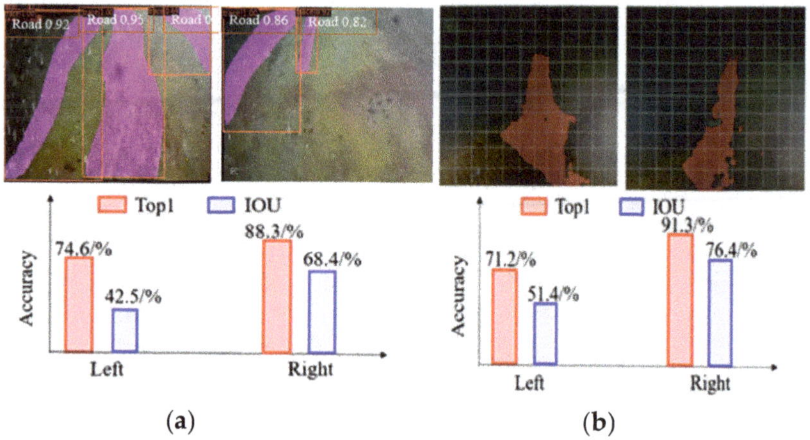

Figure 7. The segmentation of scenes is evaluated with the TOP1 and IOU indexes. (**a**,**b**) The left image corresponds to the single channel method, and the right image corresponds to the cross-attention method. Purple and red are the colors of the identified objects.

4.3. The Learning Performance Evaluation

In addition, we conduct additional comparative evaluations. The evaluation metrics for comparison included the PR curve and F_1 scores, with their respective definitions provided in Equation (8). In Equation (8), *TP* represents the probability of actual positive samples being forecasted as positive, *FP* denotes the probability of actual negative samples being forecasted as positive, *FN* represents the probability of actual positive samples being forecasted as negative, and *TN* represents the probability of actual negative samples being forecasted as negative.

$$\begin{cases} Precision = \frac{TP}{(TP+FP)} \\ Recall = \frac{TP}{(TP+FN)} \\ F_1 = 2 \cdot \frac{Precision \cdot Recall}{Precision + Recal} \end{cases} \tag{8}$$

According to the definition in Equation (8), we compared the learning performance of relevant indexes, as shown in Figure 8. For the four sub-datasets, the semantic segmentation effects were individually learned, yielding comparable performance. Additionally, the four sub-datasets were integrated, and the average learning performances were recorded. Based on the average performance curve, it is evident that the P-R curve of the single image's network is entirely covered by that of the cross-attention network, indicating the latter's superior performance.

Moreover, we can observe that the equilibrium point value (P = R) of the cross-attention network surpasses that of the single-channel network, indicating improved segmentation performance by the cross-attention network. Additionally, the cross-attention network exhibits a higher F1 score, leading us to conclude that the cross-attention network achieves state-of-the-art performance.

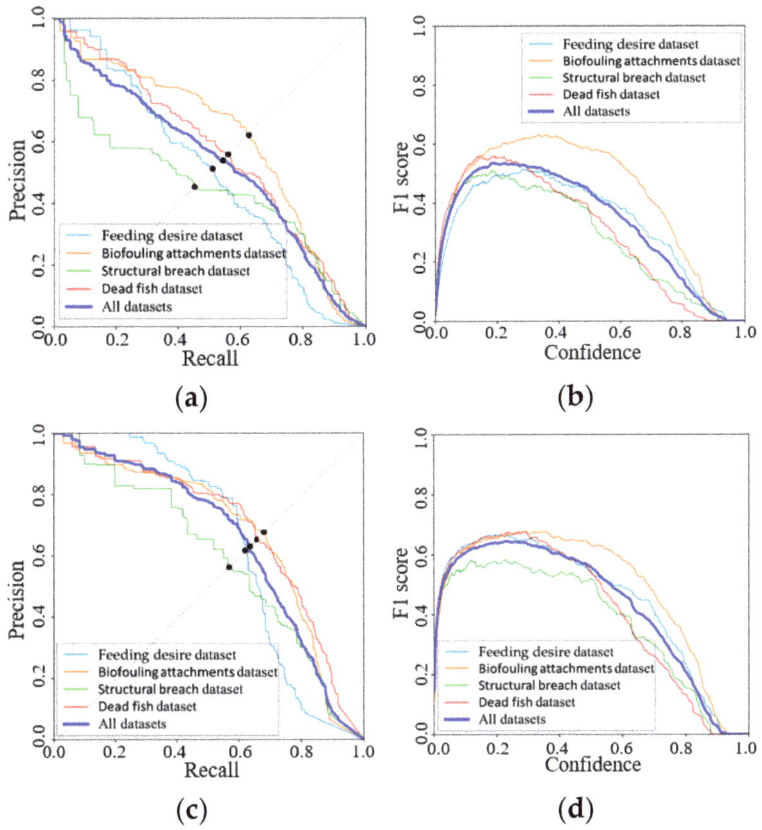

Figure 8. Comparison of detection with P-R curves and F1 score indexes. (**a**) The P-R curve for the single-channel method; (**b**) the F1 score for the single-channel method; (**c**) the P-R curve for the cross-attention method; (**d**) the F1 score for the cross-attention method.

4.4. Comparisons with Other Methods

To validate the effectiveness of our proposed components, we conducted systematic experiments by incrementally adding each module to a baseline model. The baseline consists of a standard Swin-Transformer backbone processing only RGB images. All experiments were performed on our aquaculture dataset under identical training conditions.

Table 1 presents a component-wise ablation study analyzing the incremental performance improvements of various model modifications. The baseline model, utilizing only RGB inputs, achieves 72.1% mAP and 63.5% IOU with a latency of 1.8 ms. Incorporating SNN filtering yields a performance gain of +5.2% mAP and +4.3% IOU while introducing only a marginal latency increase of 0.1 ms. The addition of cross-attention mechanisms further enhances performance, improving mAP and IOU by +6.3% and +5.4%, respectively, with a modest latency overhead of 0.2 ms. Finally, element-wise multiplication optimization provides additional gains of +4.3% mAP and +5.0% IOU while reducing latency by 0.7 ms, resulting in a final performance of 87.9% mAP and 78.2% IOU at 1.4 ms latency. Each modification demonstrates consistent improvements in both accuracy and computational efficiency.

Table 1. Component-wise ablation analysis.

Model Configuration	mAP (%)	ΔmAP	IOU (%)	ΔIOU	Latency (ms)	ΔLatency
Baseline (RGB only)	72.1	-	63.5	-	1.8	-
+SNN Filtering	77.3	+5.2	67.8	+4.3	1.9	+0.1
+Cross-Attention	83.6	+6.3	73.2	+5.4	2.1	+0.2
+Element-wise Multiplication	87.9	+4.3	78.2	+5.0	1.4	−0.7

Table 2 presents the calculation latency for various Transformer-based architectures. The first-row method, utilizing dot-product multiplication, achieved a detection accuracy of 75.1% with a latency of 1.3 ms. The second-row method replaced the dot-product multiplication with element-wise multiplication, resulting in a slightly lower detection accuracy of 74.8%, but appearing with a reduced latency of 1.1 ms. The third-row method employed the cross-attention network, yielding a performance improvement of 88.3% at the cost of a 1.6 ms computational latency. Finally, the fourth-row method combined the cross-attention network and element-wise multiplication, demonstrating commendable performance in both detection accuracy and speed. These comparative results highlight the advantages of our proposed model in terms of both accuracy and latency.

Table 2. The detection latency and TOP1 accuracy with different Transformer-based structures.

Methods	Image	Latency (ms)	Top1 (%)
Dot-product + Single channel	RGB	1.3	75.1
Element-wise + Single channel	RGB	1.1	74.8
Dot-product + Cross-attention	RGB + NeuroVI	1.6	88.3
Element-wise + Cross-attention	RGB + NeuroVI	1.4	87.9

The objects' box-detection with different backbones is compared in Table 3. Mask RCNN is used as the detection head. Our cross-attention network has achieved a performance of 42.1 AP boxes, which overcomes the AlexNet (↑9.7), VGGNet (↑8.6), UNET++ (↑5.9), and ResNet50 (↑3.1). As shown in the literature [25], we also used the Swin-Transformer method for comparison. Furthermore, our proposed cross-attention network outperforms the popular Swin-transformer, with a performance improvement of 1.9 AP boxes. In terms of instance segmentation, our method achieves a detection performance of 38.2 AP^{mask}, which is better than the previous algorithms. The cross-attention network outperforms lightweight models like AlexNet (↑9.4), VGGNet (↑8.2), Unet++ (↑4.7), and ResNet50 (↑3.8).

Table 3. The object detection and semantic segmentation with different backbones.

Backbones	Detection and Instance Segmentation (%)					
	AP^{box}	AP^{box}_{50}	AP^{box}_{75}	AP^{mask}	AP^{mask}_{50}	AP^{mask}_{75}
AlexNet [34]	32.4	52.7	35.0	28.8	50.2	30.9
VGGNet [35]	33.5	53.3	36.1	30.0	51.7	32.5
UNET++ [22]	36.2	56.1	38.9	33.5	53.6	34.5
ResNet18 [36]	35.0	55.0	37.7	32.2	52.0	33.7
ResNet50 [36]	38.0	58.6	41.4	34.4	55.1	36.7
Swin-Transformer [25]	40.3	60.5	43.6	36.3	57.4	38.4
Ours	42.1	62.2	45.1	38.2	59.8	40.2

The improvement in detection and segmentation performance demonstrates the effectiveness of our proposed backbone model. The observed enhancements in detection and segmentation performance ensure the perception safety under adverse conditions.

Prior research in underwater image segmentation has primarily relied on unimodal approaches utilizing either RGB data or sonar imaging alone. Such approaches exhibit limitations in addressing challenging underwater conditions such as suspended sediments and light scattering. To overcome these constraints, we propose NeuroVI-RGB Fusion, a novel dual-modal framework that effectively combines the complementary strengths of both visual modalities. Our architecture incorporates an innovative cross-attention mechanism that distinguishes itself from conventional Transformer-based approaches (e.g., Swin-Transformer) through its guided attention module, which dynamically enhances salient features from NeuroVI to optimize RGB segmentation performance. Furthermore, the proposed design demonstrates superior computational efficiency, achieving faster inference times (1.4 ms) compared to standard dot-product transformer implementations (1.6 ms), while maintaining competitive segmentation accuracy through its lightweight architecture.

5. Conclusions

Underwater pose significant challenges to fishery aquaculture cage tasks, as they can impact sensor and camera performance. This study proposes a dual-image cross-attention network, which can improve the cable segmentation performance under low illumination. We have pioneered the development of four low-light datasets focused on feeding desire, biofouling attachments, fence breach, and dead fishes. Secondly, we have designed a dual-channel feature extraction network, in which NeuroVI images provide cross-attention for RGB images. Finally, to reduce the detection latency, an element-wise multiplication mechanism is designed to accelerate computations among different matrices. The final qualitative and quantitative comparison demonstrate that our network has achieved the state-of-the-art detection performance.

The current approach has several limitations, including dependence on event camera availability, computational overhead for real-time deployment, and challenges with regard to generalizing to diverse aquatic environments, compounded by high GPU memory usage during training (16 GB for 1000 images). To address these issues, future work will focus on developing lightweight SNN architectures to reduce computational demands, integrating acoustic sensors for deeper water applications where event cameras are less effective, and expanding the dataset to include more species and varied environmental conditions to improve generalization.

Author Contributions: Conceptualization, Z.X.; methodology, X.L.; software, F.M.; validation, X.L. and F.M.; data curation, X.L. and F.M.; writing—original draft preparation, X.L., F.M. and Z.X.; writing—review and editing, X.L., F.M. and Z.X.; project administration, Z.X. All authors have read and agreed to the published version of the manuscript.

Funding: This work was funded in part by Central Public-interest Scientific Institution Basal Research Fund, CAFS (NO. 2023TD87), in part by the Shanghai Pujiang Program (NO. 24PJA160), in part by the Central Public-interest Scientific Institution Basal Research Fund, ECSFR, CAFS (NO. 2025YJS201).

Data Availability Statement: The original contributions presented in the study are included in the article material; further inquiries can be directed to the corresponding author.

Acknowledgments: The discussion and assistance are from the Fishery Machinery and Instrument Research Institute. Xiangyong Liu and Zhiqiang Xu are greatly appreciated. All authors listed have made a substantial, direct, and intellectual contribution to the work and approved it for publication.

Conflicts of Interest: The authors declare no conflict of interest.

References

1. Huan, X.; Shan, J.; Han, L.; Song, H. Research on the efficacy and effect assessment of deep-sea aquaculture policies in China: Quantitative analysis of policy texts based on the period 2004–2022. *Mar. Policy* **2024**, *160*, 105963. [CrossRef]
2. Sun, X.; Hu, L.; Fan, D.; Wang, H.; Yang, Z.; Guo, Z. Sediment Resuspension Accelerates the Recycling of Terrestrial Organic Carbon at a Large River-Coastal Ocean Interface. *Glob. Biogeochem. Cycles* **2024**, *38*, e2024GB008213. [CrossRef]
3. Xiao, Y.; Huang, L.; Zhang, S.; Bi, C.; You, X.; He, S.; Guan, J. Feeding behavior quantification and recognition for intelligent fish farming application: A review. *Appl. Anim. Behav. Sci.* **2025**, *285*, 106588. [CrossRef]
4. López-Barajas, S.; Sanz, P.J.; Marín-Prades, R.; Gómez-Espinosa, A.; González-García, J.; Echagüe, J. Inspection operations and hole detection in fish net cages through a hybrid underwater intervention system using deep learning techniques. *J. Mar. Sci. Eng.* **2024**, *12*, 80. [CrossRef]
5. Zhu, G.; Li, M.; Hu, J.; Xu, L.; Sun, J.; Li, D.; Dong, C.; Huang, X.; Hu, Y. An Experimental Study on Estimating the Quantity of Fish in Cages Based on Image Sonar. *J. Mar. Sci. Eng.* **2024**, *12*, 1047. [CrossRef]
6. Guan, M.; Cheng, Y.; Li, Q.; Wang, C.; Fang, X.; Yu, J. An Effective Method for Submarine Buried Pipeline Detection via Multi-sensor Data Fusion. *IEEE Access* **2019**, *7*, 125300–125309. [CrossRef]
7. Li, D.; Du, L. Recent advances of deep learning algorithms for aquacultural machine vision systems with emphasis on fish. *Artif. Intell. Rev.* **2022**, *55*, 4077–4116. [CrossRef]
8. Kong, H.; Wu, J.; Liang, X.; Xie, Y.; Qu, B.; Yu, H. Conceptual validation of high-precision fish feeding behavior recognition using semantic segmentation and real-time temporal variance analysis for aquaculture. *Biomimetics* **2024**, *9*, 730. [CrossRef]
9. Gao, T.; Jin, J.; Xu, X. Study on detection image processing method of offshore cage. *J. Phys. Conf. Ser.* **2021**, *1769*, 012070. [CrossRef]
10. Zhou, C.; Lin, K.; Xu, D.; Liu, J.; Zhang, S.; Sun, C.; Yang, X. Method for segmentation of overlapping fish images in aquaculture. *Int. J. Agric. Biol. Eng.* **2019**, *12*, 135–142. [CrossRef]
11. Hu, Z.; Cheng, L.; Yu, S.; Xu, P.; Zhang, P.; Tian, R.; Han, J. Underwater Target Detection with High Accuracy and Speed Based on YOLOv10. *J. Mar. Sci. Eng.* **2025**, *13*, 135. [CrossRef]
12. Islam, M.J.; Edge, C.; Xiao, Y.; Luo, P.; Mehtaz, M.; Morse, C.; Enan, S.S.; Sattar, J. Semantic Segmentation of Underwater Imagery: Dataset and Benchmark. In Proceedings of the 2020 IEEE/RSJ International Conference on Intelligent Robots and Systems (IROS), Las Vegas, NV, USA, 24 October 2020–24 January 2021; pp. 1769–1776.
13. Yang, D.; Wang, C.; Cheng, C.; Pan, G.; Zhang, F. Data Generation with GAN Networks for Sidescan Sonar in Semantic Segmentation Applications. *J. Mar. Sci. Eng.* **2023**, *11*, 1792. [CrossRef]
14. Liu, W.; Bai, K.; He, X.; Song, S.; Zheng, C.; Liu, X. FishGym: A High-Performance Physics-based Simulation Framework for Underwater Robot Learning. In Proceedings of the 2022 International Conference on Robotics and Automation (ICRA), Philadelphia, PA, USA, 23–27 May 2022; pp. 6268–6275.
15. Marnet, L.R.; Grasshof, S.; Brodskiy, Y.; Wąsowski, A. Bridging the Sim-to-Real GAP for Underwater Image Segmentation. In Proceedings of the OCEANS 2024—Singapore, Singapore, 15–18 April 2024; pp. 1–6.
16. Contini, M.; Illien, V.; Barde, J.; Poulain, S.; Bernard, S.; Joly, A.; Bonhommeau, S. From underwater to drone: A novel multi-scale knowledge distillation approach for coral reef monitoring. *Ecol. Inform.* **2025**, *89*, 103149. [CrossRef]
17. Liu, X.Y.; Chen, G.; Sun, X.; Knoll, A. Ground Moving Vehicle Detection and Movement Tracking Based on the Neuromorphic Vision Sensor. *IEEE Internet Things J.* **2020**, *7*, 9026–9039. [CrossRef]
18. Liu, X.Y.; Yang, Z.X.; Hou, J.; Huang, W. Dynamic Scene's Laser Localization by NeuroIV-based Moving Objects Detection and LIDAR Points Evaluation. *IEEE Trans. Geosci. Remote Sens.* **2022**, *60*, 5230414. [CrossRef]
19. Sun, X.; Chen, C.; Wang, X.; Dong, J.; Zhou, H.; Chen, S. Gaussian dynamic convolution for efficient single-image segmentation. *IEEE Trans. Circuits Syst. Video Technol.* **2021**, *32*, 2937–2948. [CrossRef]
20. Long, J.; Shelhamer, E.; Darrell, T. Fully convolutional networks for semantic segmentation. In Proceedings of the IEEE/CVF Conference on Computer Vision and Pattern Recognition 2015, Boston, MA, USA, 7–12 June 2015; pp. 3431–3440.
21. Chen, L.C.; Papandreou, G.; Kokkinos, I.; Murphy, K.; Yuille, A.L. Deeplab: Semantic image segmentation with deep convolutional nets, atrous convolution, and fully connected crfs. *IEEE Trans. Pattern Anal. Mach. Intell.* **2017**, *40*, 834–848. [CrossRef]
22. Peng, D.; Zhang, Y.; Guan, H. End-to-End Change Detection for High Resolution Satellite Images Using Improved UNet++. *Remote Sens.* **2019**, *11*, 1382. [CrossRef]
23. Huang, Z.; Wang, C.; Wang, X.; Liu, W.; Wang, J. Semantic image segmentation by scale-adaptive networks. *IEEE Trans. Image Process.* **2020**, *29*, 2066–2077. [CrossRef] [PubMed]
24. Deng, X.; Wang, P.; Lian, X.; Newsam, S. Nightlab: A duallevel architecture with hardness detection for segmentation at night. In Proceedings of the IEEE/CVF Conference on Computer Vision and Pattern Recognition 2022, New Orleans, LA, USA, 18–24 June 2022; pp. 938–948.

25. Liu, Z.; Lin, Y.; Cao, Y.; Hu, H.; Wei, Y.; Zhang, Z.; Lin, S.; Guo, B. Swin transformer: Hierarchical vision transformer using shifted windows. In Proceedings of the IEEE/CVF International Conference on Computer Vision 2021, Montreal, QC, Canada, 10–17 October 2021; pp. 10012–10022.
26. Wang, W.; Chen, Z.; Yuan, X.; Guan, F. An adaptive weak light image enhancement method. In Proceedings of the Twelfth International Conference on Signal Processing Systems 2020, Shanghai, China, 6–9 November 2020.
27. Ge, H.; Ouyang, J. Underwater image segmentation via the progressive network of dual iterative complement enhancement. *Expert Syst. Appl.* **2025**, *266*, 126049. [CrossRef]
28. Li, X.; Yu, R.; Zhang, W.; Lu, H.; Zhao, W.; Hou, G.; Liang, Z. DAPNet: Dual Attention Probabilistic Network for Underwater Image Enhancement. *IEEE J. Ocean. Eng.* **2025**, *50*, 178–191. [CrossRef]
29. Jiang, J.; Xu, H.; Xu, X.; Cui, Y.; Wu, J. Transformer-Based Fused Attention Combined with CNNs for Image Classification. *Neural Process. Lett.* **2023**, *55*, 11905–11919. [CrossRef]
30. Romera, E.; Bergasa, L.M.; Yang, K.; Alvarez, J.M.; Barea, R. Bridging the day and night domain gap for semantic segmentation. In Proceedings of the 2019 IEEE Intelligent Vehicles Symposium, Paris, France, 9–12 June 2019; pp. 1312–1318.
31. Domhof, J.; Kooij, J.; Gavrila, D.M. A Joint Extrinsic Calibration Tool for Radar, Camera and Lidar. *IEEE Trans. Intell. Veh.* **2021**, *6*, 571–582. [CrossRef]
32. Chen, H.; Xu, F.; Liu, W.; Sun, D.; Liu, P.X.; Menhas, M.I.; Ahmad, B. 3D Reconstruction of Unstructured Objects Using Information from Multiple Sensors. *IEEE Sens. J.* **2021**, *21*, 26951–26963. [CrossRef]
33. Roy, S.M.; Beg, M.M.; Bhagat, S.K.; Charan, D.; Pareek, C.M.; Moulick, S.; Kim, T. Application of artificial intelligence in aquaculture—Recent developments and prospects. *Aquac. Eng.* **2025**, *111*, 102570. [CrossRef]
34. Shelhamer, E.; Long, J.; Darrell, T. Fully Convolutional Networks for Semantic Segmentation. *IEEE Trans. Pattern Anal. Mach. Intell.* **2015**, *39*, 640–651. [CrossRef] [PubMed]
35. Wan, S.; Liang, Y.; Zhang, Y. Deep convolutional neural networks for diabetic retinopathy detection by image classification. *Comput. Electr. Eng.* **2018**, *72*, 274–282. [CrossRef]
36. He, K.; Zhang, X.; Ren, S.; Sun, J. Deep Residual Learning for Image Recognition. In Proceedings of the 2016 IEEE Conference on Computer Vision and Pattern Recognition (CVPR), Las Vegas, NV, USA, 27–30 June 2016; pp. 770–778.

MDPI AG
Grosspeteranlage 5
4052 Basel
Switzerland
Tel.: +41 61 683 77 34

Journal of Marine Science and Engineering Editorial Office
E-mail: jmse@mdpi.com
www.mdpi.com/journal/jmse